IMAGES
of America

SAN GORGONIO SEARCH AND RESCUE TEAM

ON THE COVER: Deputy Willard Farquhar, Barry Lee, and Don Walters bring up the victim of an automobile accident over the side of State Highway 38 near Angelus Oaks, California. Several times each year, cars plummet off the steep sides of mountain highways and the team is called to rescue the victims and unfortunately, sometimes, recover the bodies of the deceased. (Photograph courtesy of Don Walters.)

IMAGES
of America

SAN GORGONIO SEARCH AND RESCUE TEAM

Bob Lehmann and Bob Blanck

ARCADIA
PUBLISHING

ISBN 978-0-7385-5576-8

Published by Arcadia Publishing
Charleston SC, Chicago IL, Portsmouth NH, San Francisco CA

Printed in the United States of America

Library of Congress Catalog Card Number: 2007930847

For all general information contact Arcadia Publishing at:
Telephone 843-853-2070
Fax 843-853-0044
E-mail sales@arcadiapublishing.com
For customer service and orders:
Toll-Free 1-888-313-2665

Visit us on the Internet at www.arcadiapublishing.com

In memory of Phil Calvert and Scott Johnston, who were killed in the line of duty, and the brave men and women who voluntarily take time away from their careers and loved ones to search for the lost and rescue the injured without compensation or reward both in San Bernardino County and around the country.

Contents

ACKNOWLEDGMENTS

Special thanks to Rod Walters as the team's unofficial historian with his library of historical and recent photographs. To Don Walters, who has served on the team since the beginning and who helped to identify the people and places in the early photographs and provide a valuable oral history of the team. Recognition needs to be given to all of our teammates who do this job in the worst of weather and at all hours of the day and night at their own personal expense. The risk to health and well-being can be great; some team members have been injured while others have made the ultimate sacrifice in the line of duty. We volunteer knowing the dangers and will go out again when the call sounds.

Special recognition also needs to be made to the spouses (especially Jennifer Blanck, who put in many long hours helping edit captions and compiling pictures for this book) and the families of the team members who hope and pray that we return safely. Without their love and support, we could not do what we do so well.

Unless otherwise noted, all photographs used in this book belong to members of the San Gorgonio Search and Rescue Team.

INTRODUCTION

"YOU HAVE TO KNOW WHO HAS YOUR BACK and you have to trust him," became the unspoken mantra of the newly formed San Gorgonio Search and Rescue Team.

In February 1958, a group of students and their parents traveled to Forest Falls, California, to play in Mill Creek. One of the students, Donald Burns, wandered off and fell over 300 feet down the face of Big Falls near the picnic area where his friends and family played. Big Falls is one of the tallest waterfalls in Southern California, dropping over 400 feet in a series of cascades and falls. Big Falls is very accessible to anyone venturing into the Forest Falls area, and a short trail from the Falls picnic area (once a campground) leads to the base of the falls. There is a concrete pathway that leads to the lower pool with a very obvious sign warning everyone who may want to climb the falls that they are risking their lives. There is a large rock near one of the pools that was called "Blood Rock" by the locals because it is where many of the victims of climbing accidents come to their final rest. Burns miraculously survived his fall but was left stranded and severely injured on a ledge over 100 feet above the base of the falls. Deputy Sheriff Willard Farquhar of the San Bernardino County Sheriff's Department was the first to respond to the scene, and as night enveloped the falls, Burns could be heard screaming over the roar of the water.

Attempts to reach Burns by Deputy Farquhar and other sheriff's personnel were unsuccessful, and San Bernardino County had no experienced rescue personnel or climbing equipment available. Volunteer firemen from Forest Falls and some rescue personnel from Norton Air Force Base also made attempts to save Donald. They worked for 24 hours straight to reach Burns but were unable to get closer than about 50 feet. It was said that one rescuer could see Burns feebly waving his hand to him, but all attempts at rescue were unsuccessful. Conditions that night were cold and foggy. The ropes were freezing with the spray of water from the falls, and the rocks in the area were coated with a thick layer of ice.

The next day, the sheriff's office called for a U.S. Marine Corps helicopter. The marines lowered a cable down to Burns but were unable to get close enough to rescue the boy. A call was finally made to the closest mountain search-and-rescue team in the area, the Sierra Madre Search and Rescue Team, which was associated with the Los Angeles County Sheriff's Department. This team was requested to respond and attempt a rescue. Seven members responded to the scene, escorted by Sheriff Frank Bland and the California Highway Patrol. After a briefing by the sheriff about where the boy was located and the conditions on Big Falls with the water spray, ice, and crumbling rock, six members started up the falls. They climbed approximately 400 vertical feet along the west side of the falls through the brush and forest to a point where they could secure their ropes. One member descended his rope near the vertical falls to a point 50 feet above the boy. From there, he was lowered in a free rappel the remaining distance to Burns. When asked, "Is he alive?" over the walkie-talkie, the reply was "Negative." Donald was dead, covered with a thick layer of ice. He had more than likely died from a combination of exposure to the cold and his injuries sometime during the night.

The boy was recovered, his body strapped into a basket and raised up the falls; but the rope kept getting caught in rock crevices, making the recovery process very slow and difficult. It was getting dark, so a huge searchlight was brought in and used to light up the mountainside. The basket was eventually raised to the anchor point of the main line above the falls. Due to darkness and the exhaustion of the rescuers, it was decided that it was too risky to complete the recovery that night. The basket with Burns's body was left on the mountainside to be carried out by sheriff's personnel the following morning.

After the tragic death of Donald Burns, Sheriff Frank Bland never wanted such a thing to happen in "his" county again. Sheriff Bland, being in charge of the sheriff's office for the largest county in the state, did not like going outside of his county for services. Knowing that Deputy Willard Farquhar had some experience in hiking and climbing in the local mountains, the sheriff gave him the task of forming the county's first mountain search-and-rescue team.

In March 1958, Deputy Farquhar started recruiting men from the local area who had experience in the mountains. Some individuals had military experience with rope rescue, others had backcountry experience with the U.S. Forest Service, and still others had a strong desire to help their fellow man in the challenging situations and environments the wilderness offers. The San Bernardino County Board of Supervisors at the time thought that a mountain search-and-rescue team would be valuable to the county, not only for searching for the lost and rescuing the injured but also for educating the public about the dangers of the mountains and what was needed to be prepared when venturing into the wilderness. The board of supervisors pledged that they would assist with monetary support for equipment and the maintenance of vehicles.

In March 1958, the first volunteer members were given their depositions. But it was not until May 1958 that they were officially sworn in by Sheriff Bland. By the time of the "swearing-in" ceremony, the team already had three missions under their belts: the recovery of a skier who was killed in an avalanche on Mount Baldy, the recovery of two bodies from a plane crash on Cucamonga Peak, and the rescue of two injured firefighters.

Because of the high number of accidents occurring in the Big Falls area of Forest Falls, in June 1958, Deputy Farquhar formed the Valley of the Falls Search and Rescue Team as a quick response or "hasty" team to handle calls in the area. The initial roster for the Valley of the Falls team had 17 members; it was required that they all be Forest Falls residents and their missions confined to the area of Forest Falls and Mill Creek Canyon. The new Valley of the Falls Team members were sworn in on June 23, 1958. As "sister" teams, the Valley of the Falls and San Gorgonio Search and Rescue Teams have worked closely together for many years, supporting and assisting each other to this very day.

Fifty years later, the San Gorgonio Search and Rescue Team still rushes to the aid of the lost and the injured in the San Gorgonio Wilderness and the San Bernardino National Forest.

One

Out of Tragedy is Born a Team

As a result of the death of Donald Burns on Big Falls in February 1958, Sheriff Frank Bland appointed Deputy Willard Farquhar to form a mountain search-and-rescue team. In March 1958, Deputy Farquhar recruited men from the local area who had experience in the mountains. It was decided that the team would be strictly volunteer.

It was up to the team to raise money for equipment and vehicles from the local communities. The cities of Redlands, Yucaipa, and unincorporated Mentone were small towns with a lot of community spirit and were willing to help the newly formed team. Presentations were made to local community service organizations, and articles were written in the *Redlands Daily Facts* in support of the team's activities and the need for donations. A benefit dinner was held at the Mentone School cafeteria; over 400 people were served. The first team fund-raiser was a huge success. In the first year, two military surplus Jeeps, a Jeep trailer, and team rescue equipment were purchased.

In 1959, the team added a new member, Trooper, a seven-month-old bloodhound. Purchased by the team, Trooper was the first bloodhound in San Bernardino County trained for wilderness search-and-rescue work. This purchase represented a considerable investment by the team of donated money.

A unique event brought attention to the bloodhounds and the team. In December 1963, Trooper II and Andy Brown changed roles from the searchers to the searched for. They escaped from their kennel at the "Barn" on Ed Sampson's property in Mentone and roamed the streets of Redlands for several days, causing a media frenzy in the local newspapers and numerous sightings by the local citizens before finally being located. Both dogs had been struck by cars and injured (fortunately not seriously). After receiving a tip from a citizen, Trooper II was found by team members in Redlands. The next day, Andy Brown came home on his own to the "Barn." This was actually their second escape from the kennels. After this, a new entrance to the kennel was built to help prevent them from escaping again.

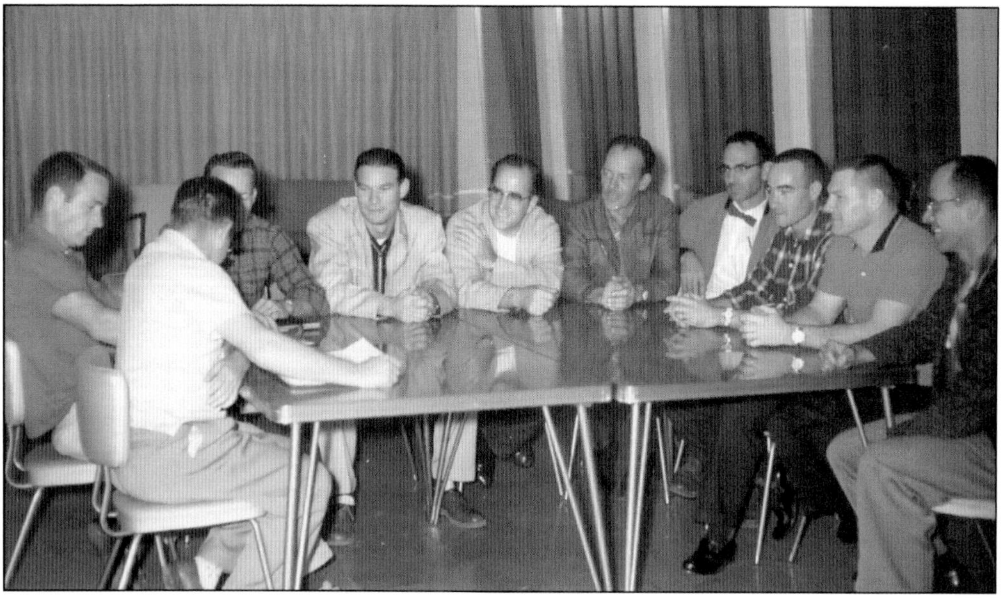

Seen here is one of the first regular business meetings of the San Gorgonio Search and Rescue Team. At this early date, there was no official meeting site, so meetings were held in one of the member's homes or at the local school. All members were from the Redlands, Mentone, and Yucaipa areas. Pictured are, from left to right, Fred Couch, first team president; Phil Farquhar; unidentifed (obscured); Ed Sampson; Dr. Barry Lee; Bob Carter; Thom Mayfield; Dave Farquhar; Don Walters; and George Gruver.

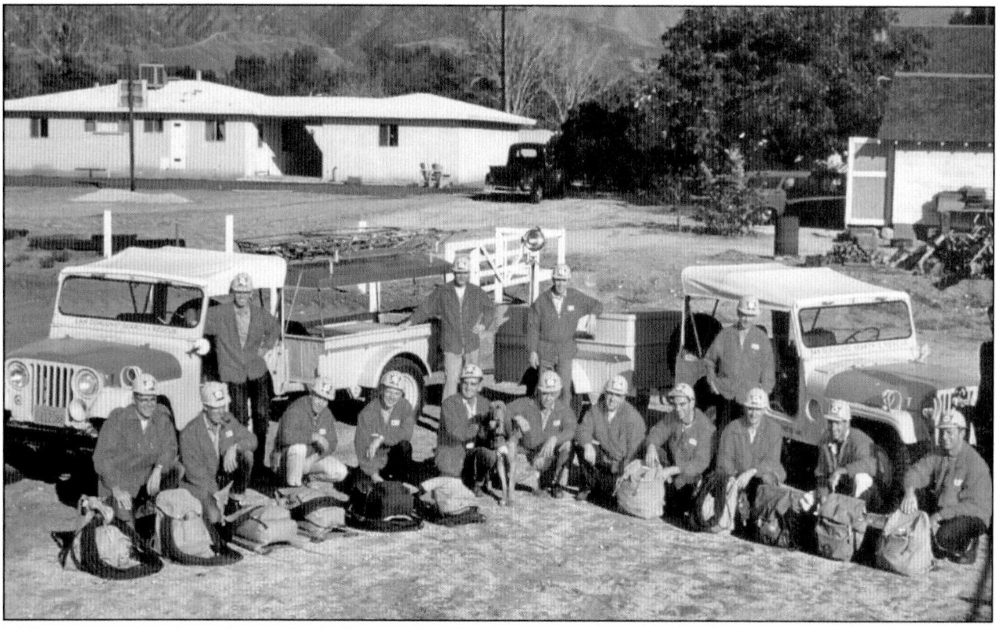

Taken in 1960, this photograph shows the team, wearing their newly issued jackets, with their "new" vehicles and equipment. The team purchased and restored these army surplus Jeeps and utility trailers. There was an obvious attempt to standardize the equipment that each team member carried. The rope is army surplus "goldline," which was the standard in 1960 for rock climbing and rescue work.

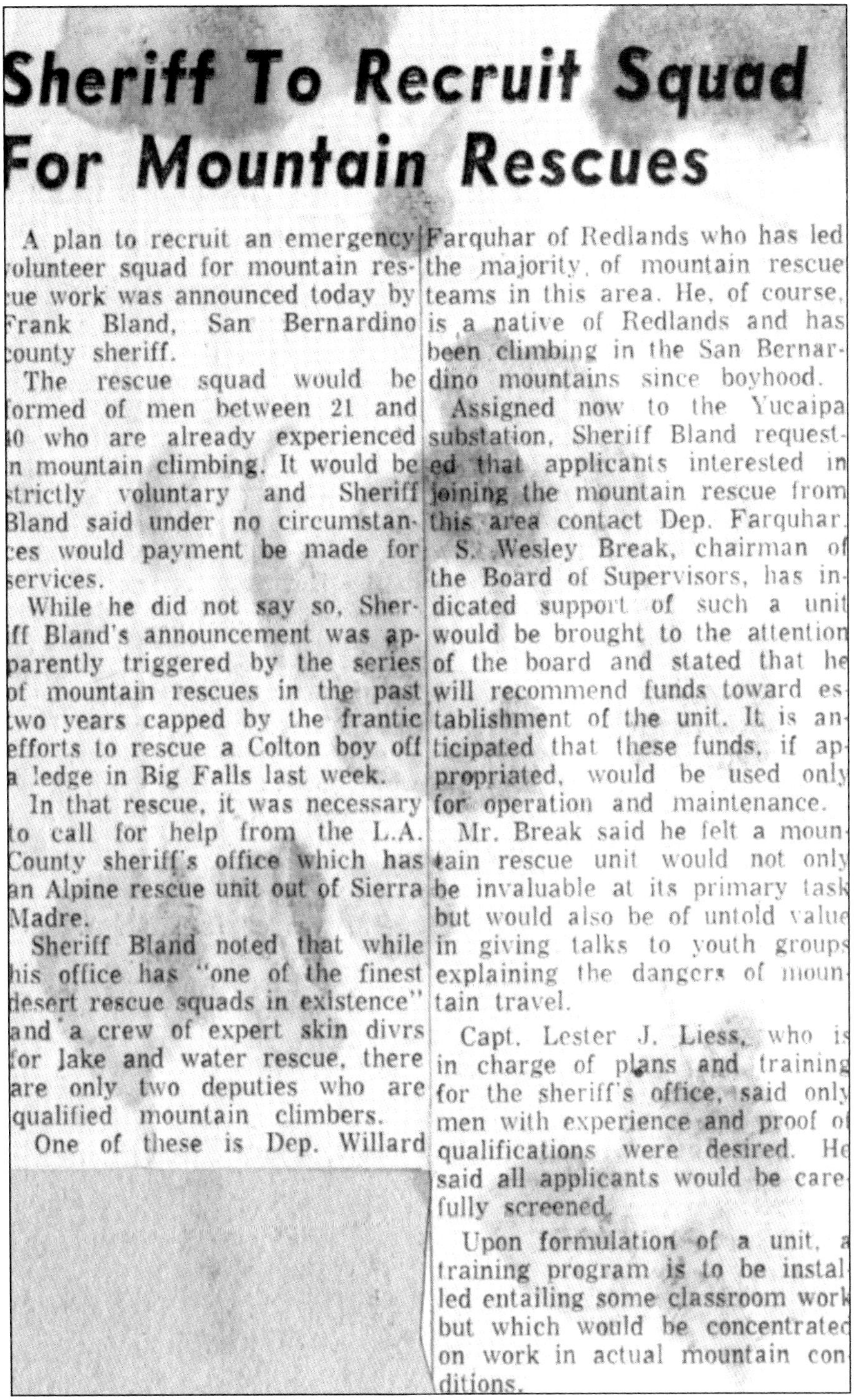

Sheriff To Recruit Squad For Mountain Rescues

A plan to recruit an emergency volunteer squad for mountain rescue work was announced today by Frank Bland, San Bernardino county sheriff.

The rescue squad would be formed of men between 21 and 40 who are already experienced in mountain climbing. It would be strictly voluntary and Sheriff Bland said under no circumstances would payment be made for services.

While he did not say so, Sheriff Bland's announcement was apparently triggered by the series of mountain rescues in the past two years capped by the frantic efforts to rescue a Colton boy off a ledge in Big Falls last week.

In that rescue, it was necessary to call for help from the L.A. County sheriff's office which has an Alpine rescue unit out of Sierra Madre.

Sheriff Bland noted that while his office has "one of the finest desert rescue squads in existence" and a crew of expert skin divrs for lake and water rescue, there are only two deputies who are qualified mountain climbers.

One of these is Dep. Willard Farquhar of Redlands who has led the majority of mountain rescue teams in this area. He, of course, is a native of Redlands and has been climbing in the San Bernardino mountains since boyhood.

Assigned now to the Yucaipa substation, Sheriff Bland requested that applicants interested in joining the mountain rescue from this area contact Dep. Farquhar.

S. Wesley Break, chairman of the Board of Supervisors, has indicated support of such a unit would be brought to the attention of the board and stated that he will recommend funds toward establishment of the unit. It is anticipated that these funds, if appropriated, would be used only for operation and maintenance.

Mr. Break said he felt a mountain rescue unit would not only be invaluable at its primary task but would also be of untold value in giving talks to youth groups explaining the dangers of mountain travel.

Capt. Lester J. Liess, who is in charge of plans and training for the sheriff's office, said only men with experience and proof of qualifications were desired. He said all applicants would be carefully screened.

Upon formulation of a unit, a training program is to be installed entailing some classroom work but which would be concentrated on work in actual mountain conditions.

In March 1958, Sheriff Frank Bland announced that the rescue team would consist of men between the ages of 21 and 40 who had prior experience with mountain climbing. All applicants were to be carefully screened for their qualifications. The start of the rescue team came as a result of the tragic death of a young boy on Big Falls the previous month.

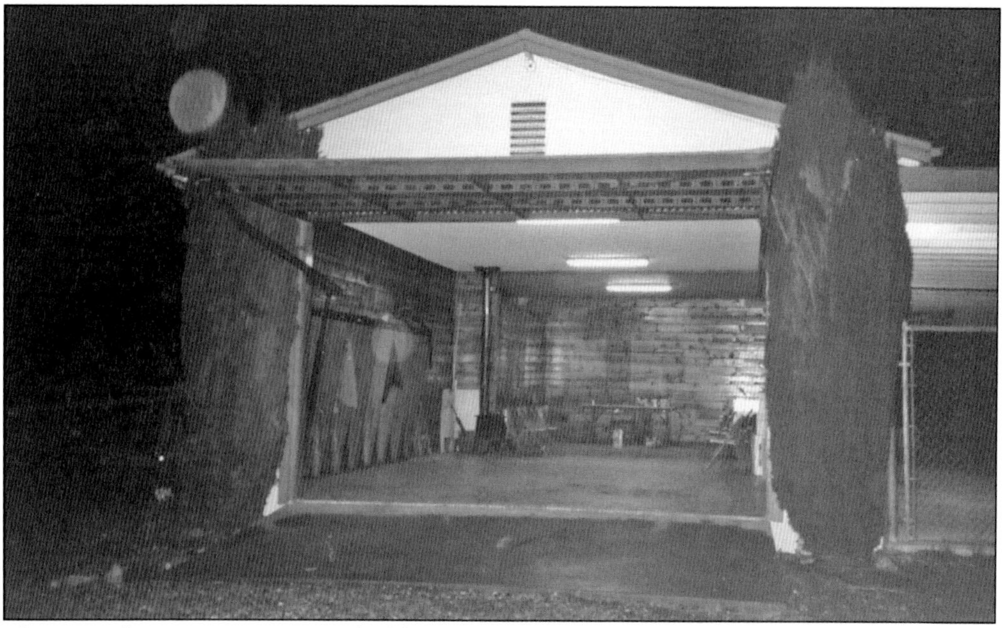

The team's first equipment storage building, affectionately known as the "Barn," was built by team members on property owned by Ed Sampson. The building could hold three to four vehicles, and wooden lockers were built to hold each team member's personal rescue equipment. Team business meetings and social events were also held in the Barn. The property is stilled owned by the Sampson family.

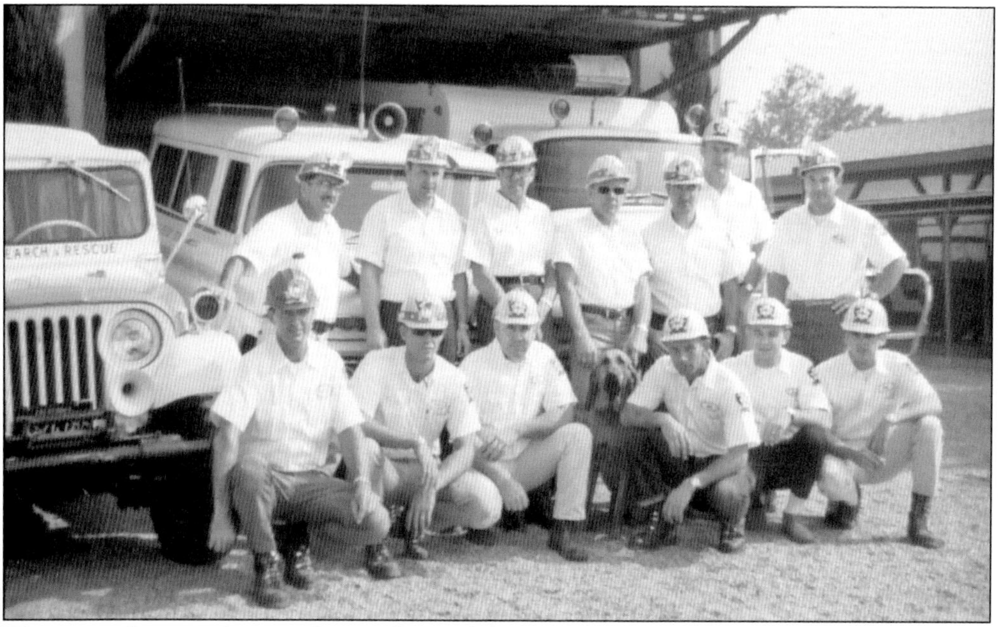

This team photograph was taken in front of the Barn in 1962. Note that uniforms had been developed—white shirts with the individual's name and team patch on the left shoulder. Pictured are, from left to right, (first row) Willard Farquhar, Dave Bushnell, Ed Walker, Andy Brown (bloodhound), Bill Loenhorst, George Gruver, and Dave Richards; (second row) Bob Carter, Ray Atzet, Thom Mayfield, Ed Sampson, Don Walters, Patrick Wassell, and Barry Lee.

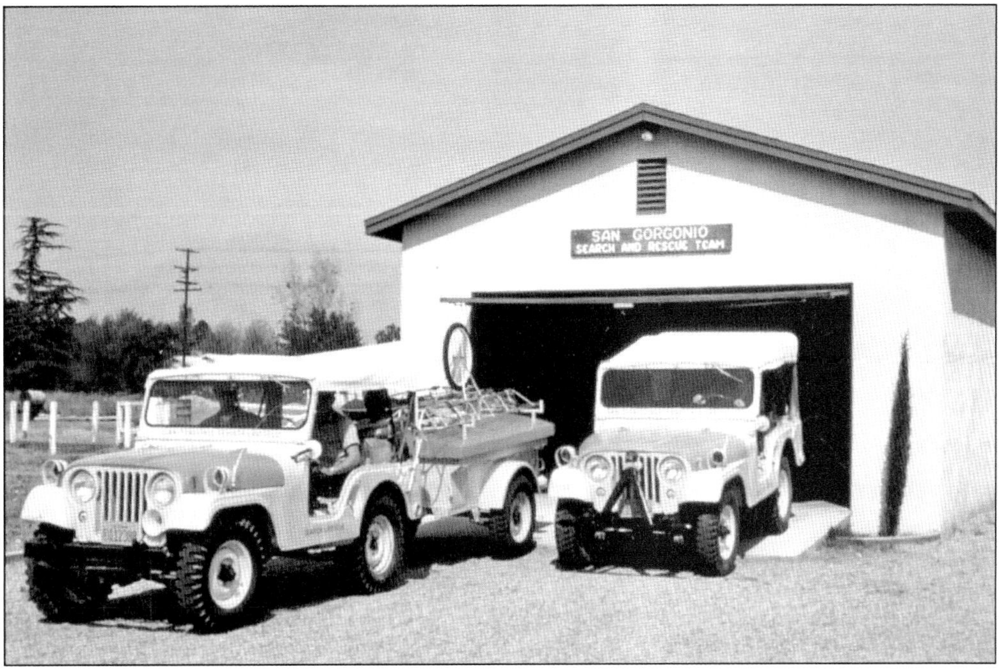

Vehicles suitable for all types of road conditions were by far the largest expenditure of funds that the team had to make in the early days. The first vehicles were surplus Jeeps from the military that were refurbished by team members for use. Yellow was the preferred color with a bright orange hood. Surplus military utility trailers were also purchased and refurbished to carry the large amount of equipment that was needed for a rescue.

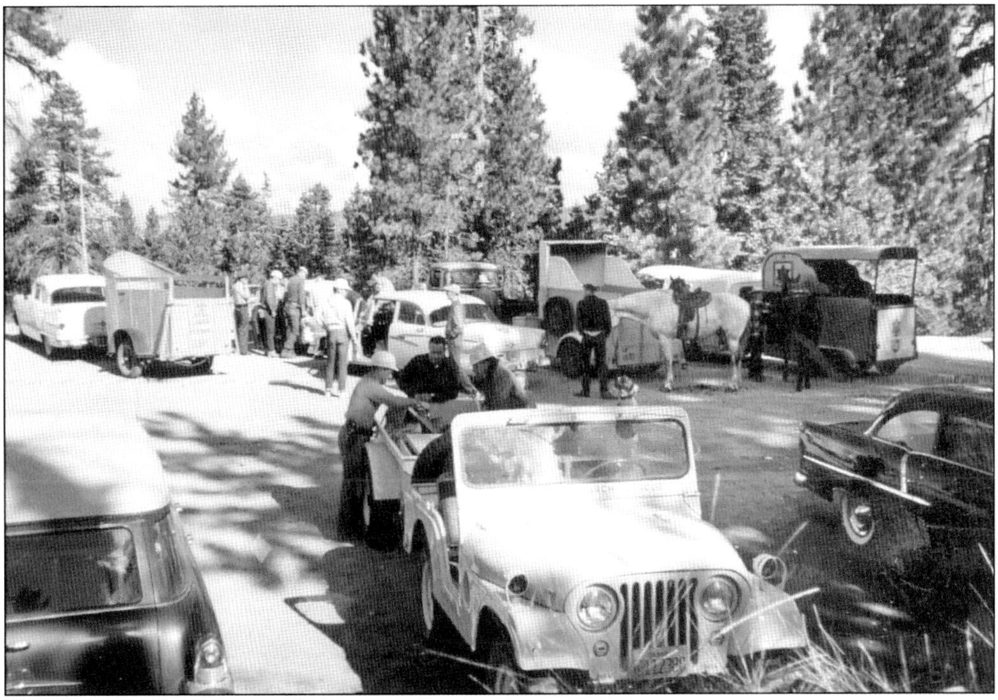

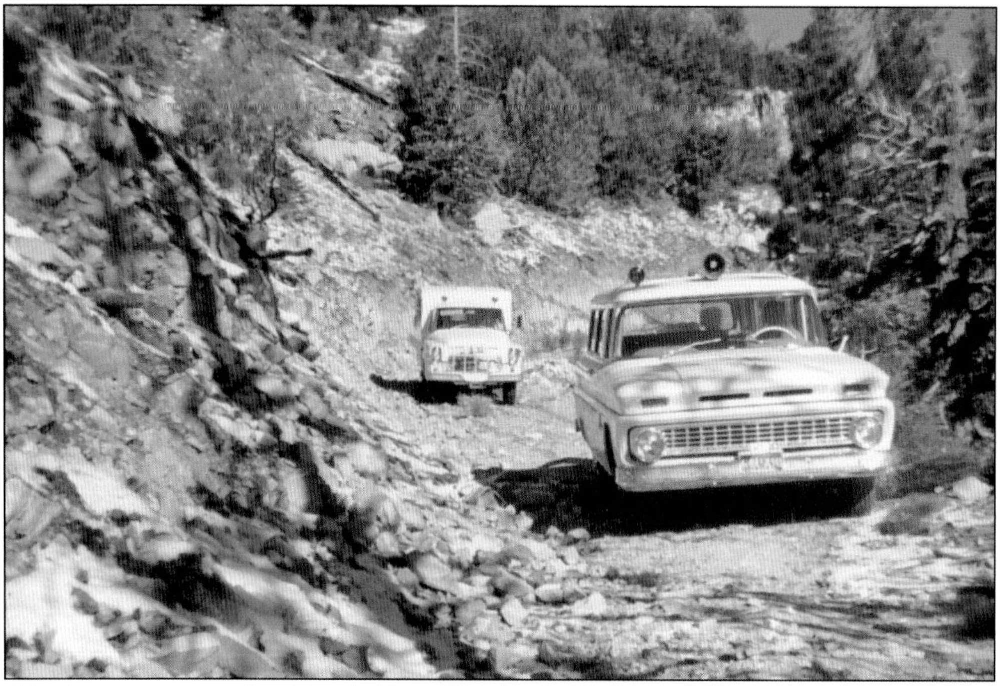

Vehicles were purchased with funds donated by the surrounding communities. The International Harvester truck, affectionately known as the "Corn Binder," was equipped with a specialized PTO power winch attached to the front bumper. This was used to lower and raise victims and personnel for over-the-side rescues. This vehicle also carried a majority of the rescue gear in the back and on the roof. The Chevrolet Carry All was used mainly as personnel transport. This vehicle is still owned by the team today, although it is only used for parades and other fund-raising events. It is the oldest operating vehicle in the sheriff's fleet.

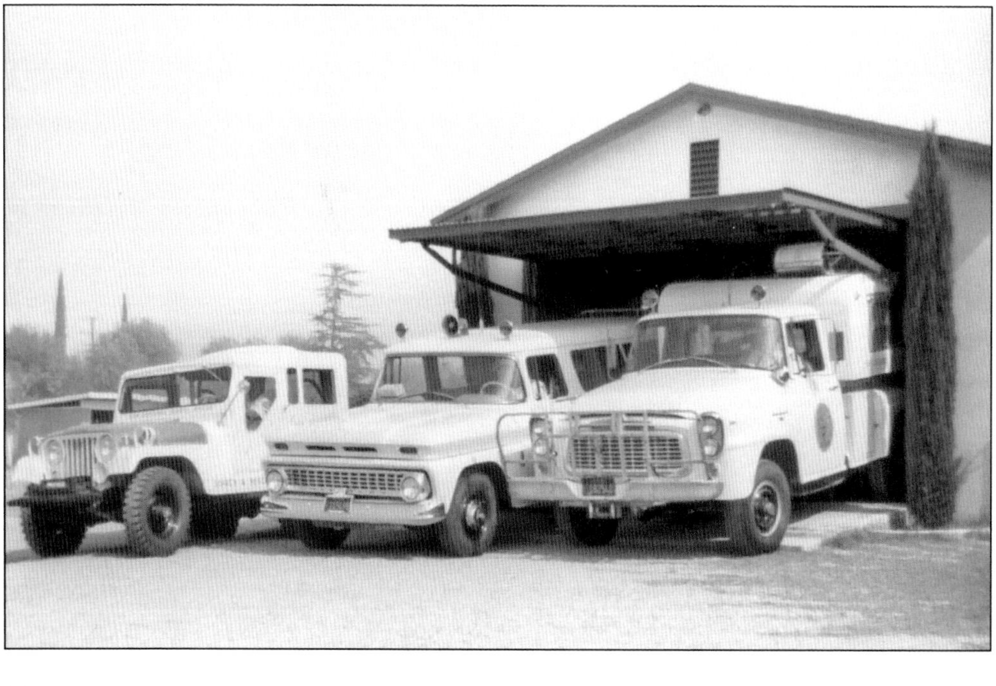

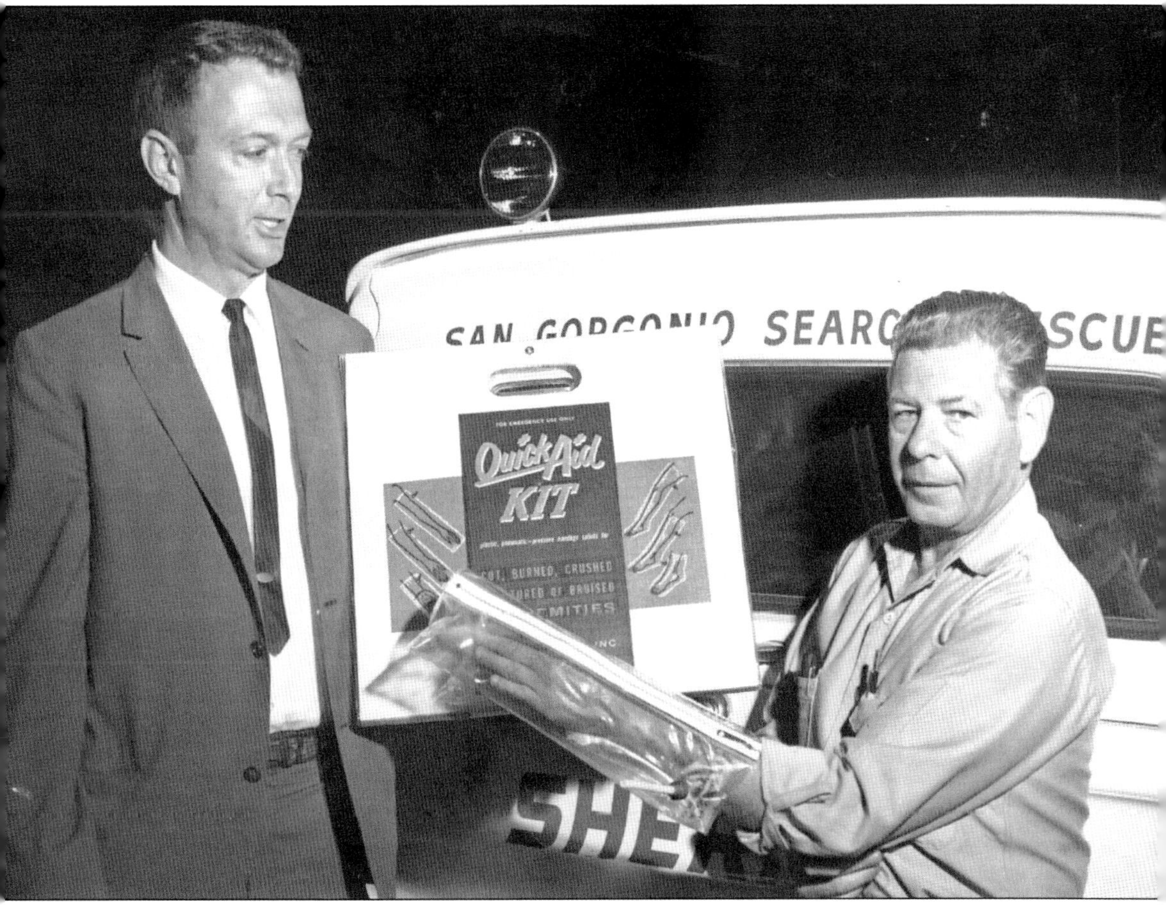

Deputy Willard Farquhar and Fred Couch demonstrate the "new" air splints, which are inflatable plastic tubes that can be placed around broken bones, burns, or even severe lacerations to protect the injury until the patient can be transported to medical care. Air splints are still used today by medics and rescue crews.

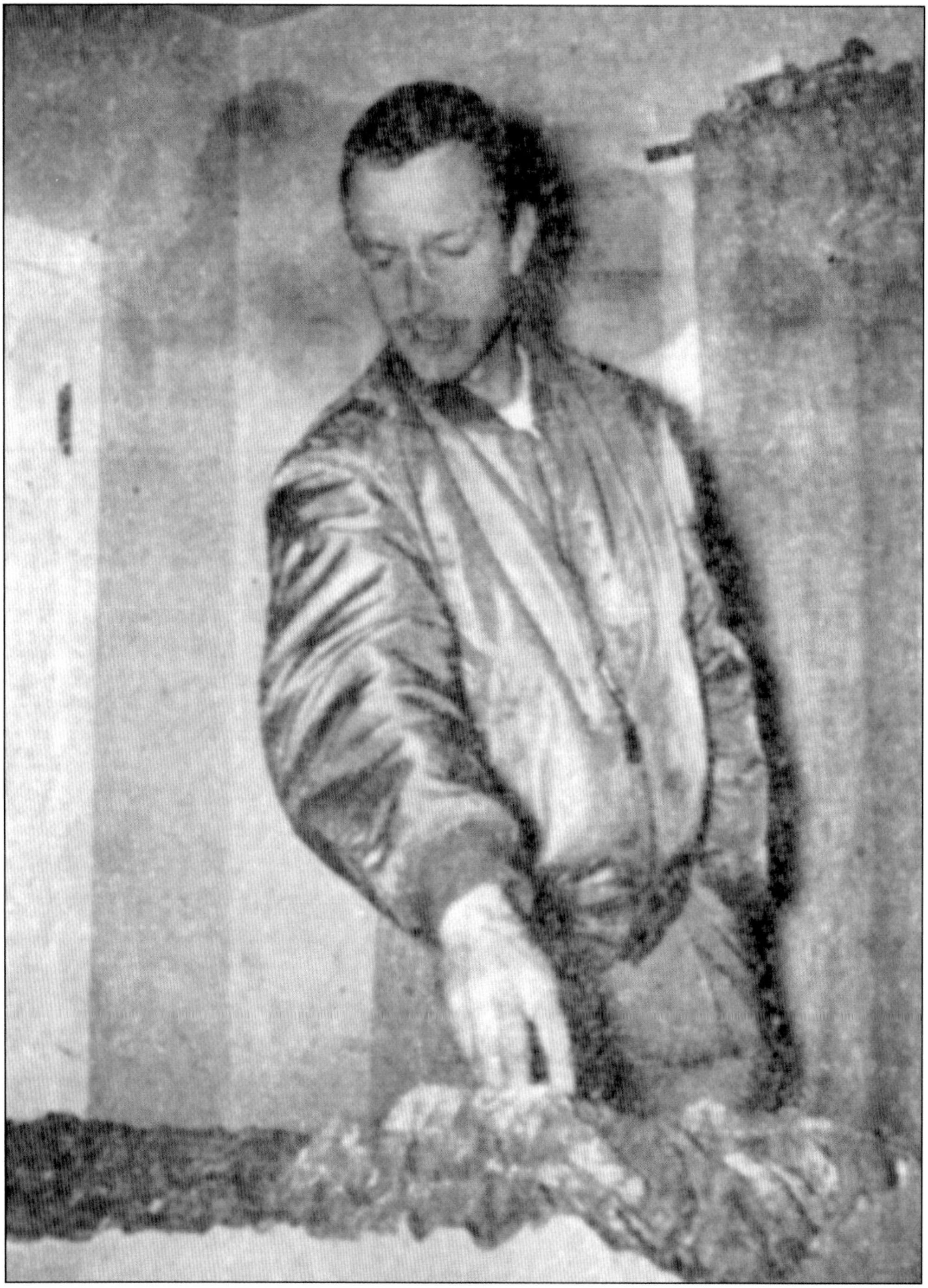

Deputy Willard Farquhar uses a relief map made by his brother, David Farquhar, to point out the areas of the San Gorgonio Wilderness to a service club in Redlands. The article quotes Willard as saying that "the San Gorgonio mountain area is as familiar to me as my own backyard." Willard was the founder and organizer of the team in 1958.

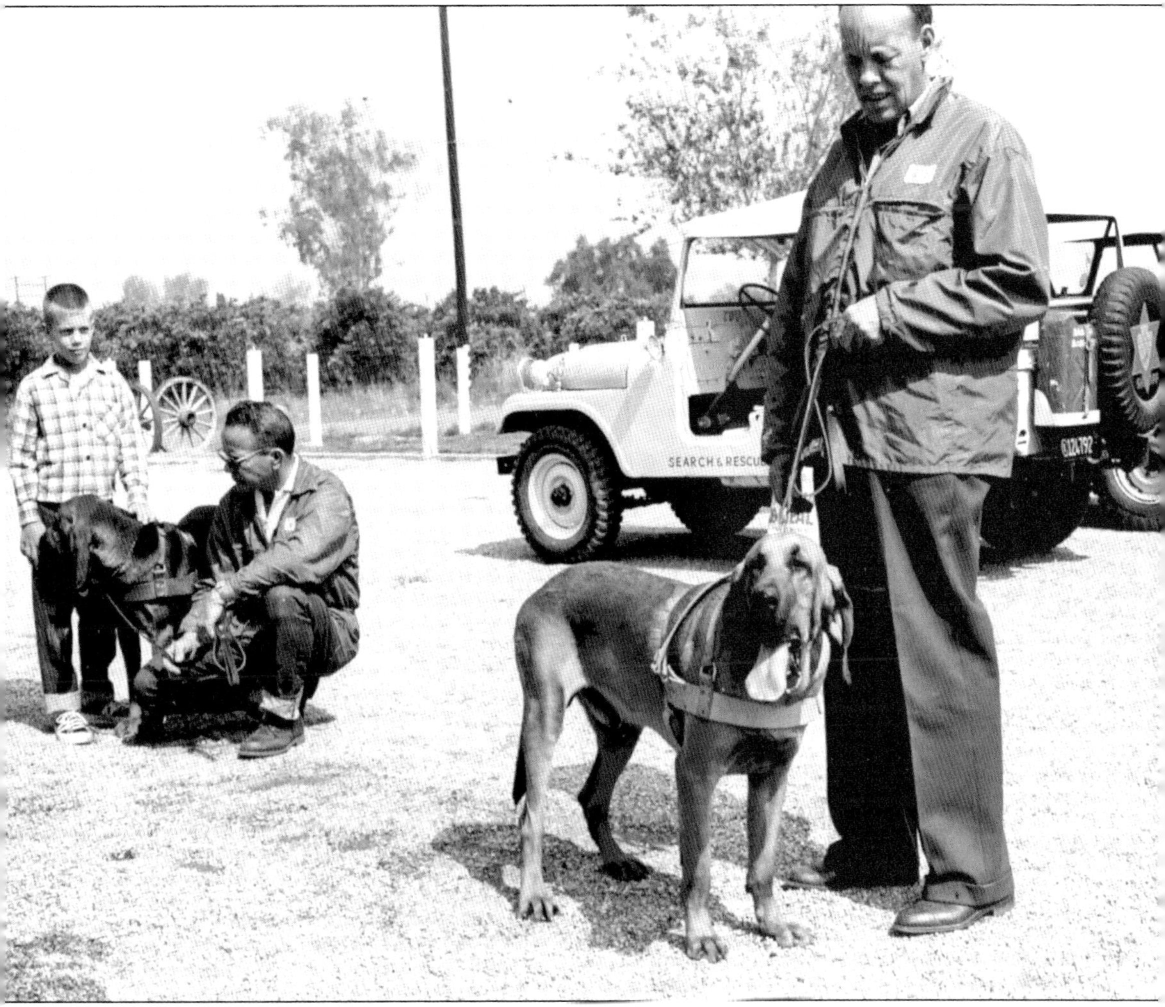

Trooper II and Andy Brown are seen with Thom Mayfield (standing) and Ed Sampson (kneeling). The dogs are wearing their rescue harnesses, which were necessary to keep them under control once they acquired the scent of the lost person.

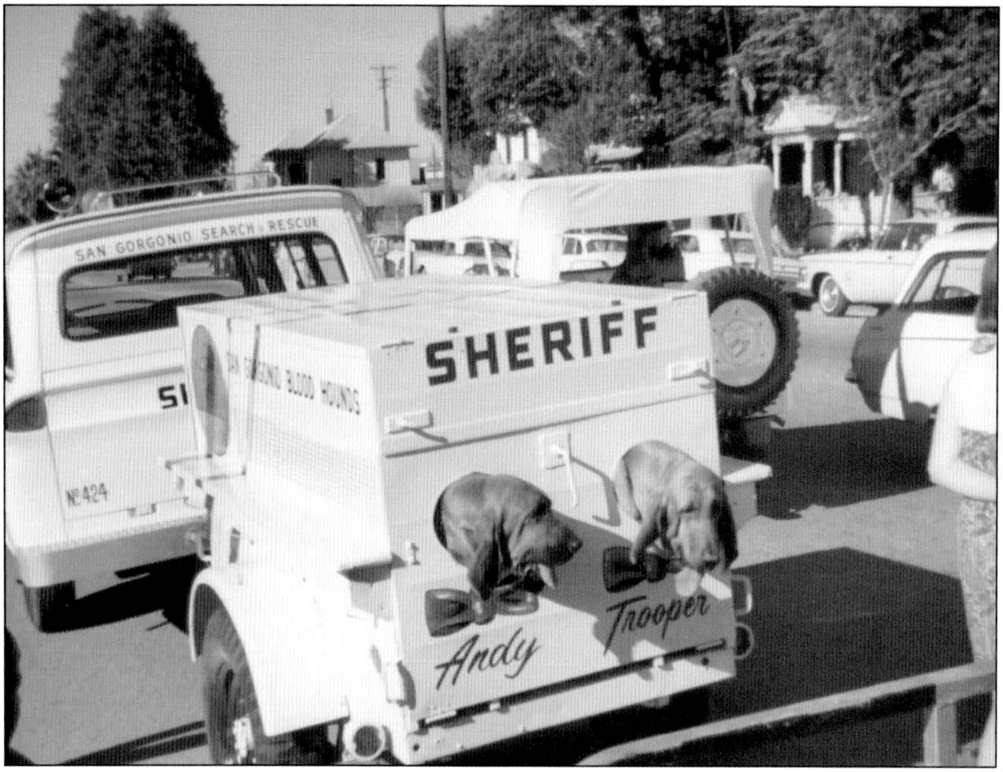

The team's first bloodhound was Trooper, a large black-and-tan male. Trooper was housed at the Barn in Mentone and was cared for by Ed Sampson. Unfortunately, Trooper died at an early age in 1961 from a liver ailment. Shortly after Trooper came Andy Brown, a black-and-tan male who was also housed at the Barn and cared for by Ed Sampson.

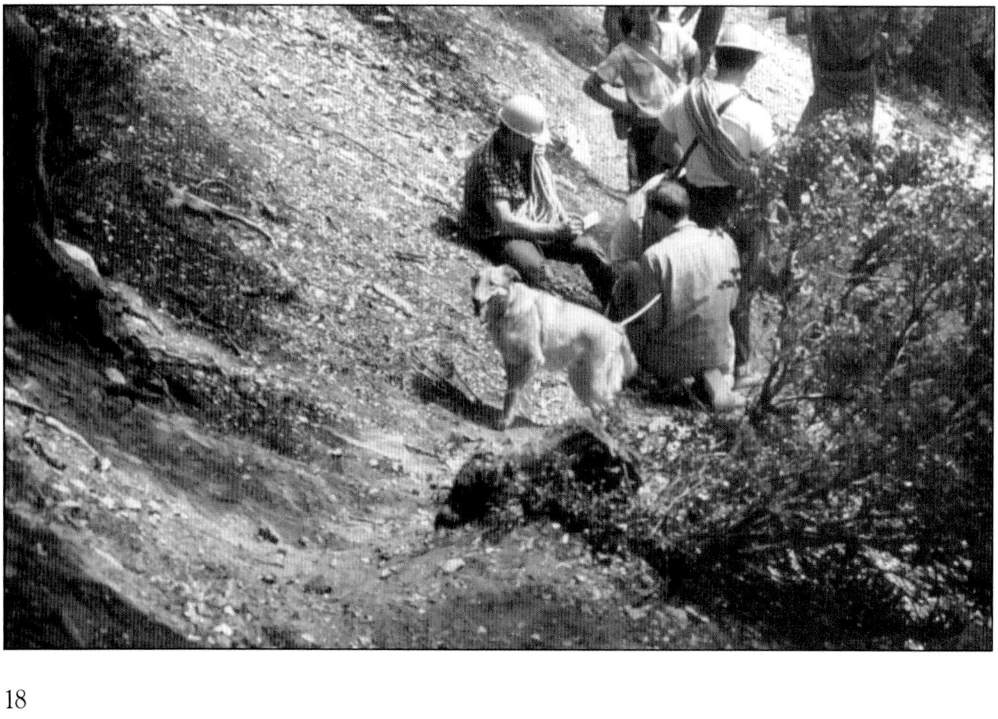

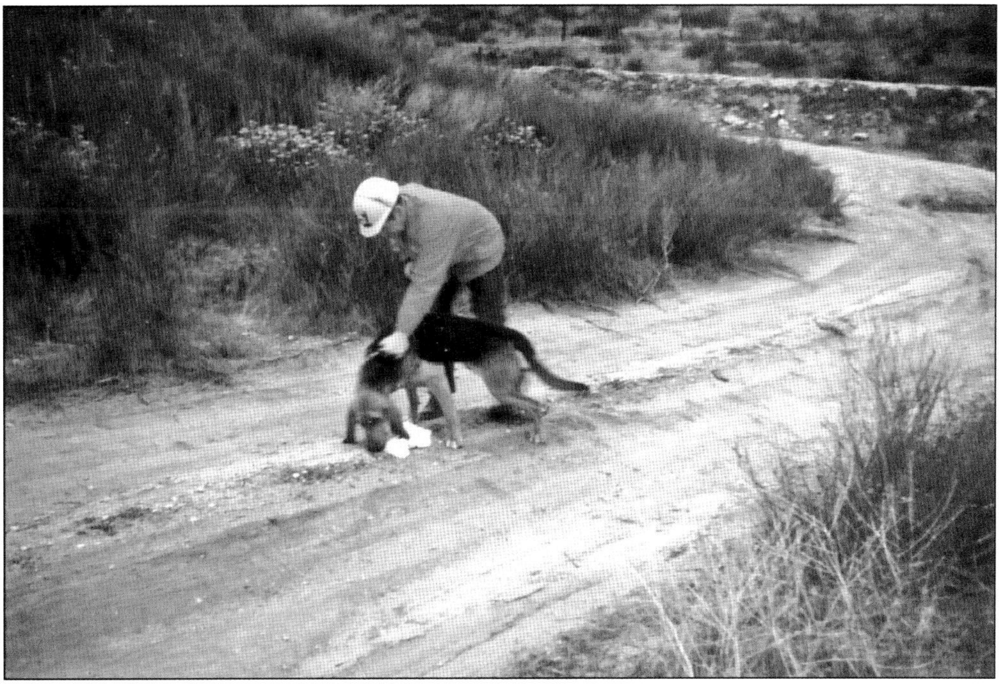

Trooper II, a red male, was a donation from the Gearhart family in Redlands. Unfortunately, Trooper II and Andy Brown died before they were able to be used to their full potential. Although they had been successful on a number of searches, the team decided that the expense and time involved in caring for and training dogs was too much. The sheriff's department would not use search dogs again regularly until the early 1990s when it started a volunteer search dog team.

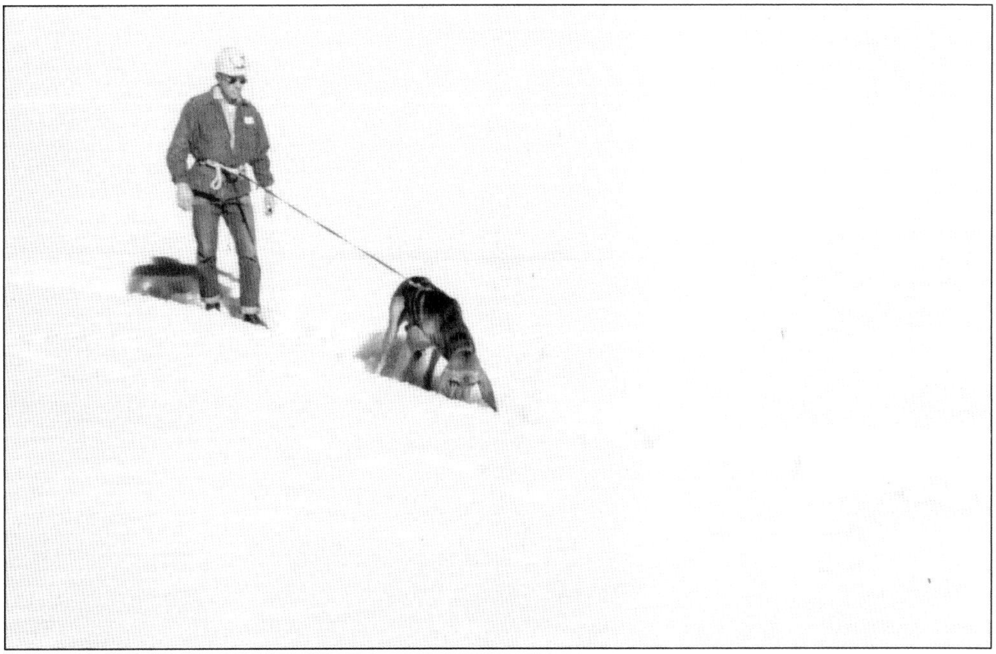

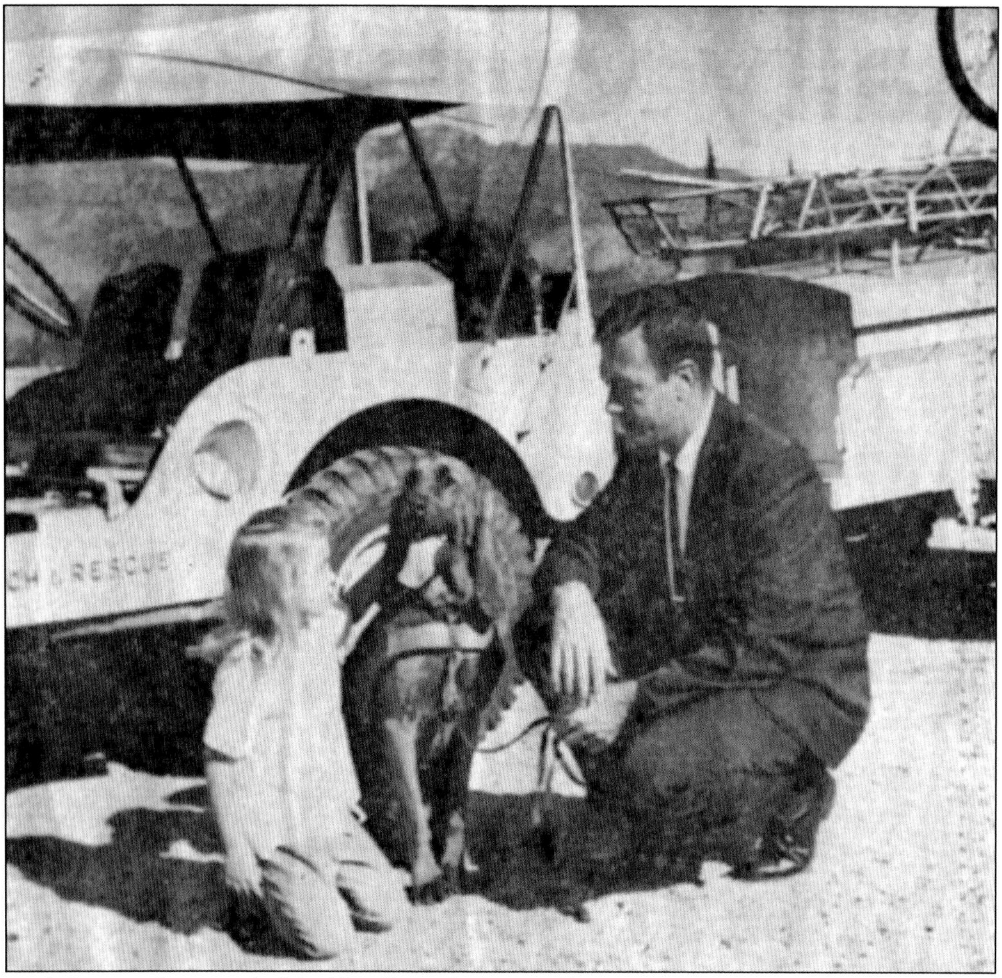

Andy Brown, the second of the bloodhounds obtained by the team, starts his training as a search dog. Trooper, the team's first bloodhound, had died from a liver ailment about a month before. Ed Sampson was the dog's trainer and keeper. Ed, who had grown very close to Trooper, would use the same training techniques on Andy Brown.

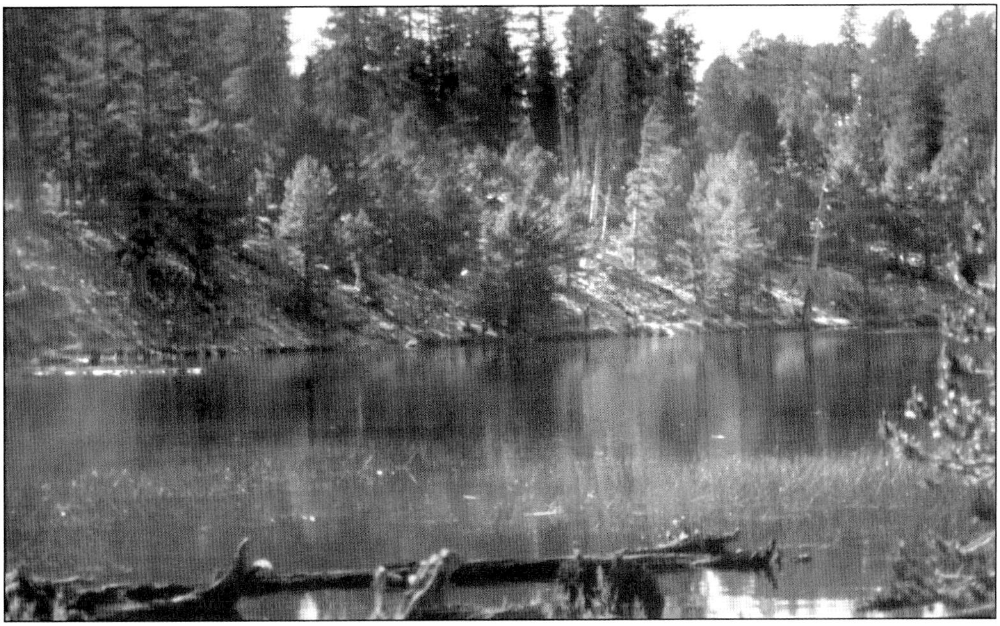

There are two dominant lakes in the San Gorgonio Wilderness, Dry Lake and Dollar Lake. Usually only having water during the wet years, Dry Lake's water levels vary and some years is even completely dry, hence its name. However, Dry Lake is an important water source for wildlife and backpackers as well as a wilderness landmark with two campsites. It is frequently used by hikers as an overnight stopping point prior to hiking to the summit of San Gorgonio.

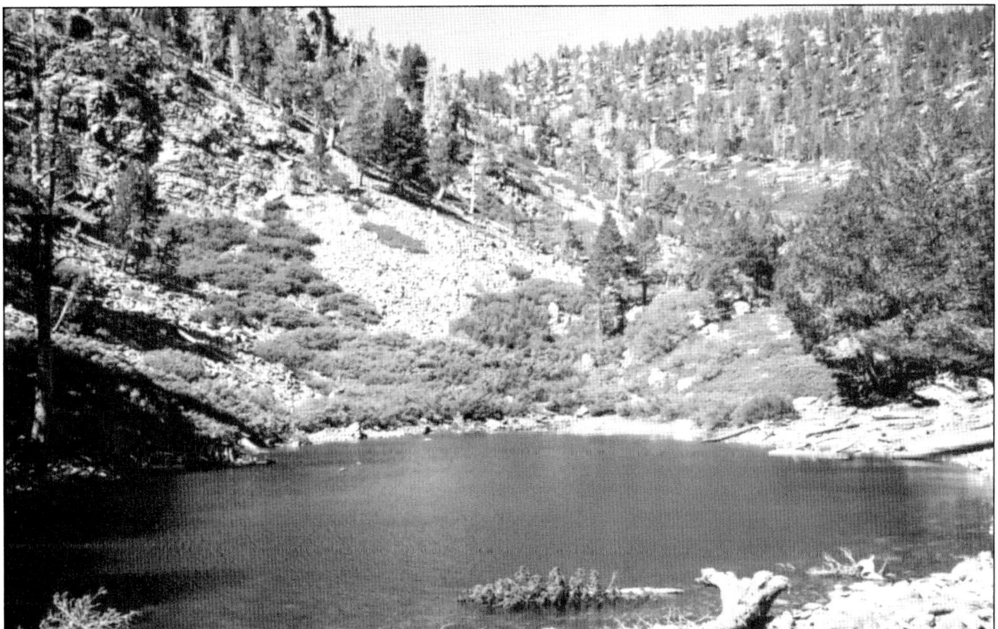

West of Dry Lake, Dollar Lake is nestled against the rocky slopes above Slushy Meadows. It is called Dollar Lake because when viewed from higher elevations, "it shines like a silver dollar." Dollar Lake also has a trail camp located about one quarter of a mile west of the lake. The springs that feed the lake and the lake itself seldom go dry.

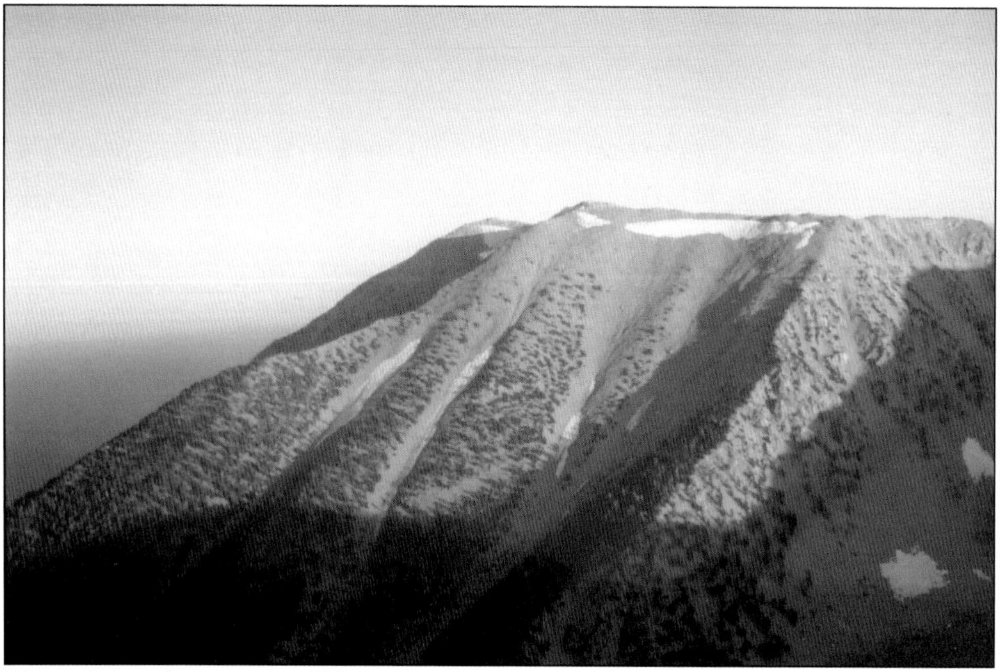

The rocky, treeless summit of Mount San Gorgonio, the highest peak in Southern California at 11,499 feet, rises above the San Gorgonio Wilderness. In wetter years, snow will remain on the north-facing slopes all year long. The views from the summit are breathtaking on clear days, looking into Mexico to the south and Catalina Island to the west. However, weather conditions at the summit can change without notice, going from clear and warm to thunderstorms, blizzards, and strong winds. The summit and the surrounding wilderness attract thousands of visitors every year. Some of these visitors make wrong turns and become disorientated and lost. Most either self rescue or become subjects of searches. A few are never found and become part of the history of the wilderness area.

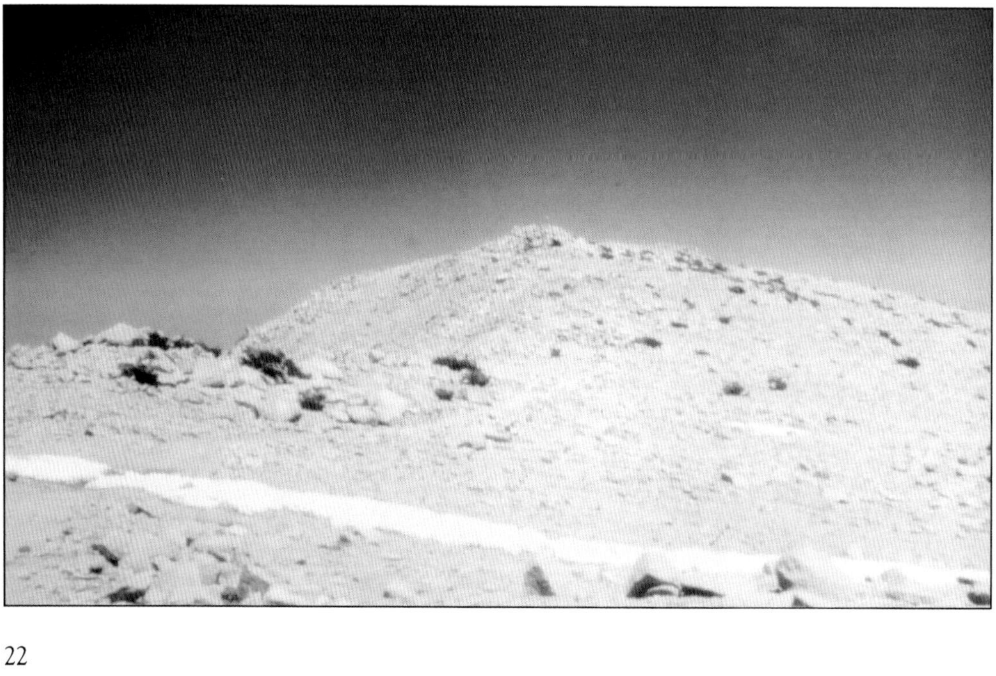

Two

The Early Years

For searches and rescues, 1958 was a very busy year. In its first nine months, the team had 19 missions, rescued 46 people, recovered 6 bodies, and logged more than 3,000 hours.

In March, the team had two body recoveries in the same week. In August, the team was called to rescue an ill scout master who was camping in the Dobb's Cabin area. Also in August, the wreckage of a light aircraft was found and the team recovered the bodies. In September, the team recovered the body of a 64-year-old man who was hunting with his 76-year-old friend in the Wilshire Peak area. The 76-year-old man, who hiked all night to report the death of his friend, was so exhausted, he slept for nearly 12 hours before he contacted authorities. Because of the man's inaccurate description of where his friend's body was, it took six days for the rescue team to find and recover the body.

A few days after the Wilshire Peak recovery, a deer hunter who lived in Highland was reported overdue by his wife. He was hunting alone in the South Fork area near Dry and Dollar Lakes. The weather did not cooperate, with heavy rains wiping out all tracks and cold nighttime temperatures not instilling much hope for survival. When the man was not found in the South Fork area, the search was expanded to the rest of the San Gorgonio Wilderness. When the hunter was still not found, the search quickly took on giant proportions with the sheriff's department calling for the marines, horses, and small aircraft. This was the largest search the sheriff's office had undertaken to date in San Bernardino County. The search was finally called off after a month as any hope of survival faded and all areas had been covered and re-covered many times. To call off a search when the subject has not been found is one of the most difficult decisions to make. Even more difficult is the duty to inform the family that all hope of finding their loved one alive is gone.

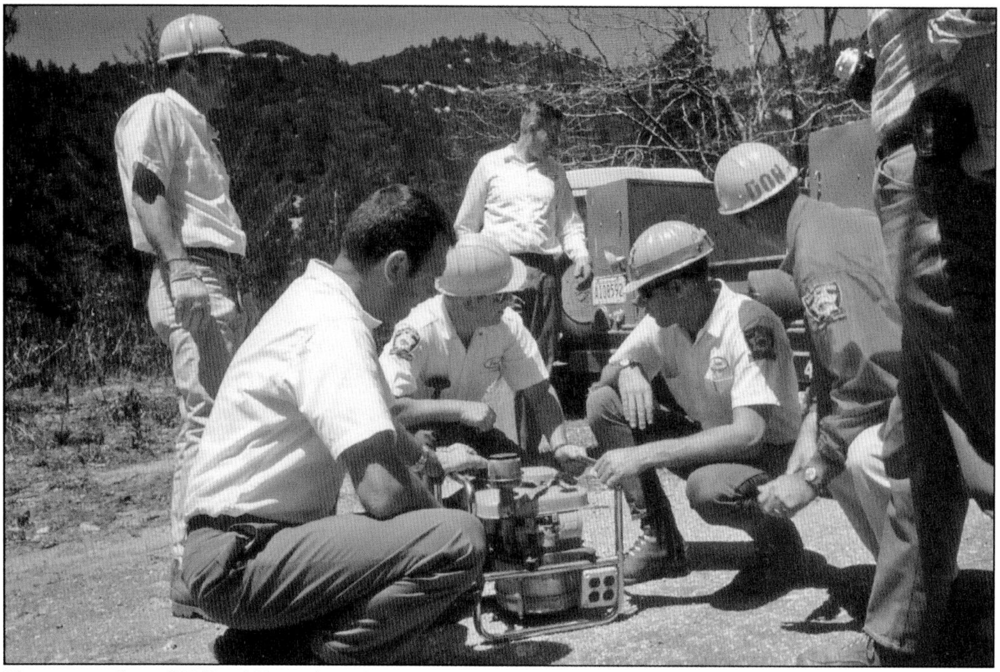

Team members Willard Farquhar, Barry Lee, Dave Bushnell, and Don Walters are inspecting, cleaning, and repairing a gasoline-powered generator donated to the team by Redlands locals. The San Gorgonio Search and Rescue Team has always been a nonprofit, volunteer organization that relies heavily on donations from the community for equipment.

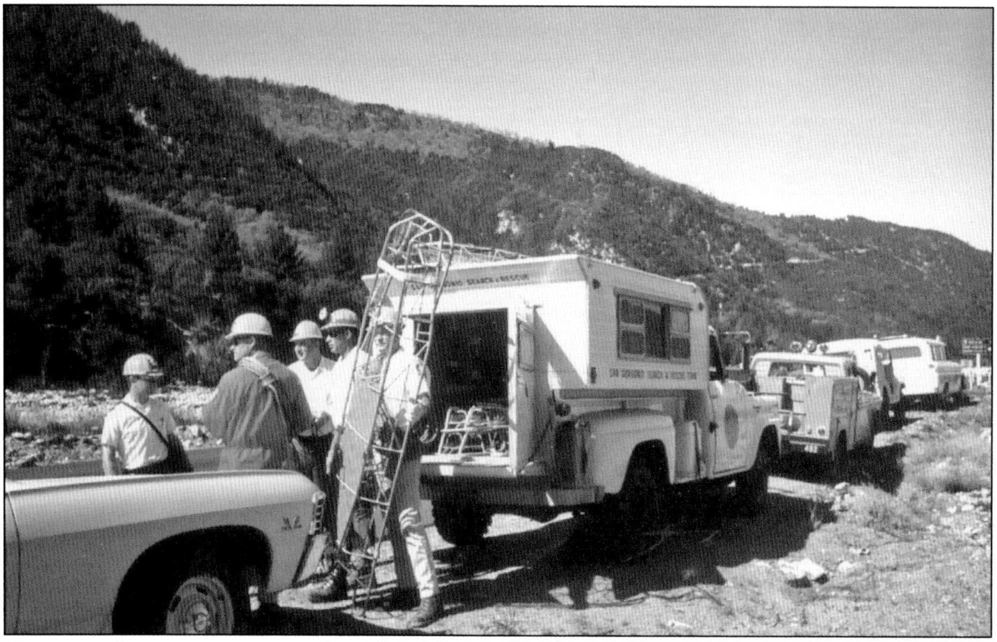

Team members are staging on the side of the road and preparing to recover the body of climber who had fallen off Monkey Face Falls in Mill Creek Canyon a few miles west of Forest Falls. Monkey Face Falls is the lesser known of the two main waterfalls in Mill Creek Canyon. It is smaller and less accessible than the larger Big Falls.

Willard Farquhar is mounting a new spotlight on a recently donated vehicle, as well as cleaning out the toolbox on the back. These first team vehicles were invaluable for storing rescue equipment and quickly getting team members to the site of injured and lost hikers.

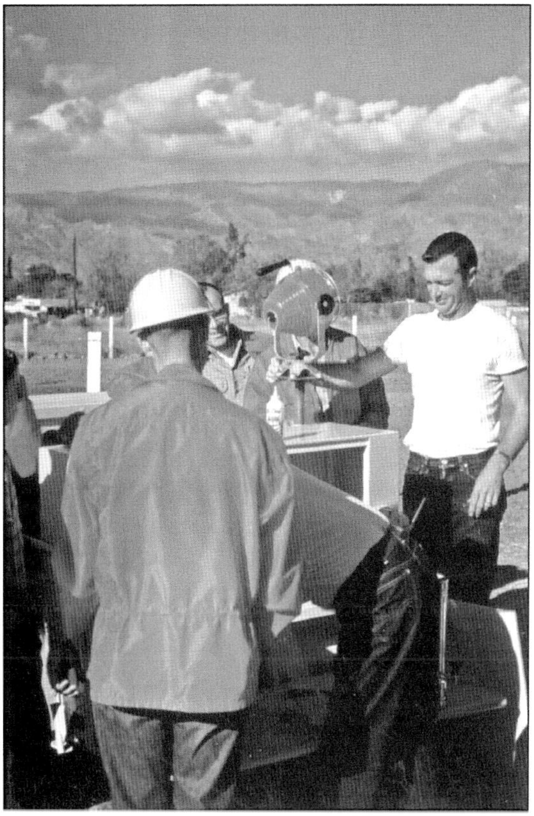

In the early days, the hood of a vehicle many times became the "command post" for rescue operations. Team members gather around maps and a "lunch box" radio to plan the search and possible rescue of a lost hiker in the San Gorgonio Wilderness.

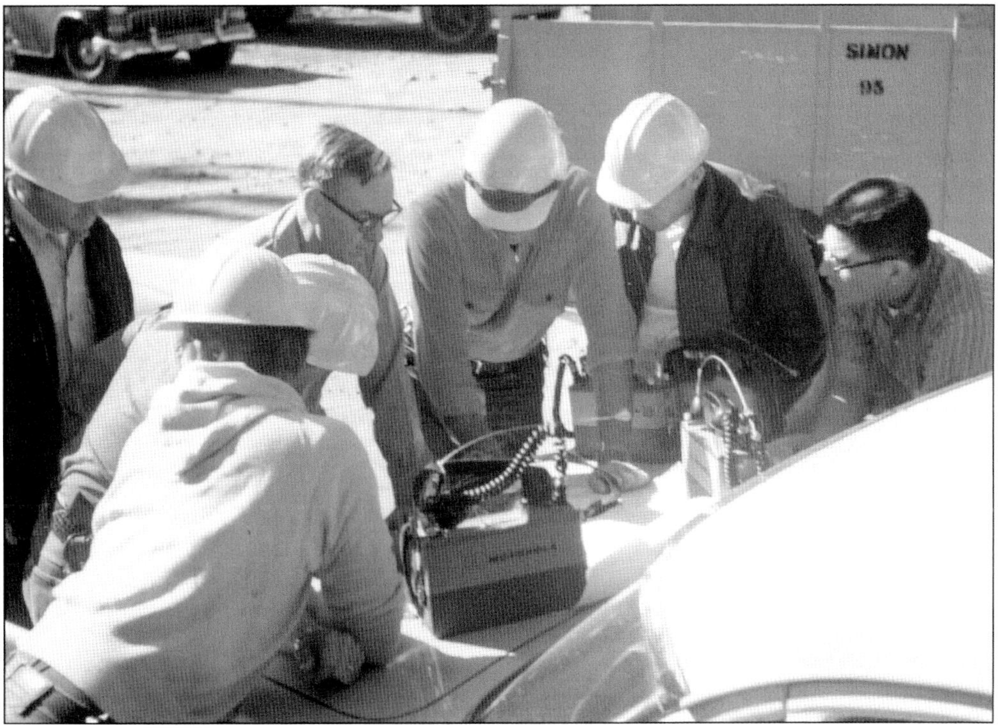

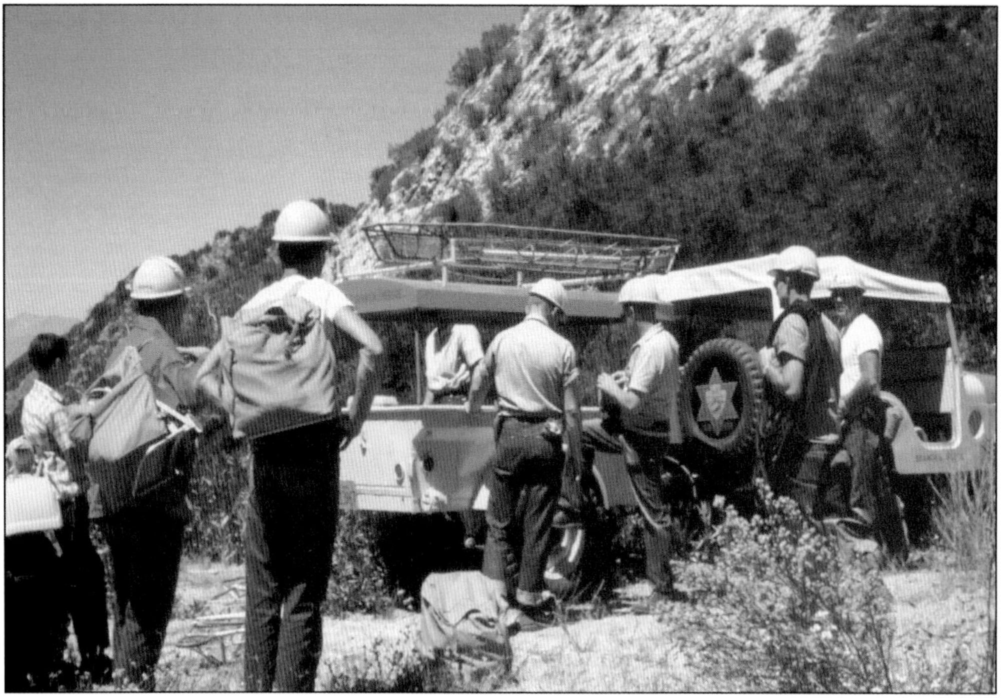

Rescue team members stage for a search and possible rescue in Mill Creek Canyon. The Jeep and trailer were among the first vehicles and equipment that the team used to get into the wilderness quickly. In these early days, the team did not yet have a standard uniform. The canvas-framed backpacks would be considered archaic by today's high-tech equipment standards but were standard issue in the late 1950s.

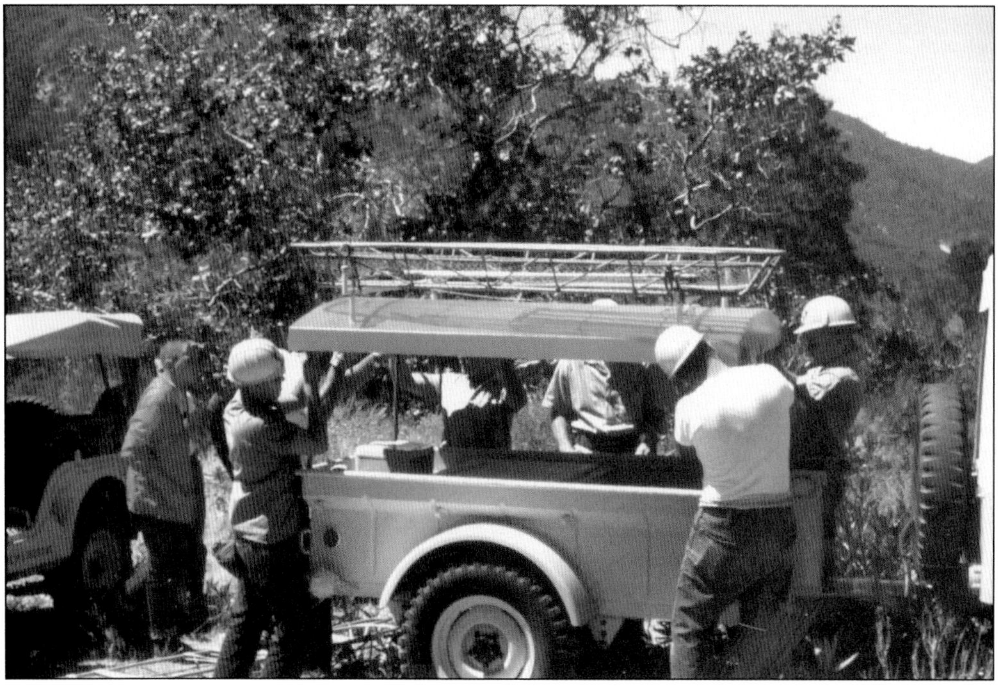

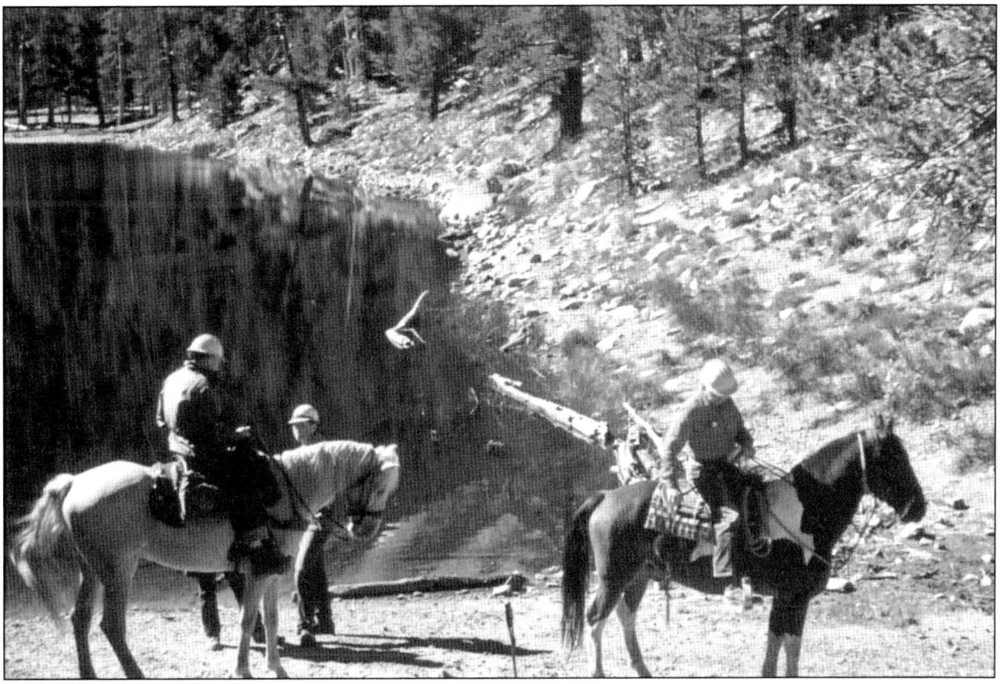

Dry Lake is one of the most beautiful spots in the San Gorgonio Wilderness. It is also one of the most frequented by hikers and therefore one of the most common places where people get lost. From there, searchers will look for traces and tracks of the lost hikers and start searching the surrounding areas. Dry Lake is great spot to camp if searchers have to spend the night in the wilderness. It also offers a large landing area for helicopters to pick up or drop off team members.

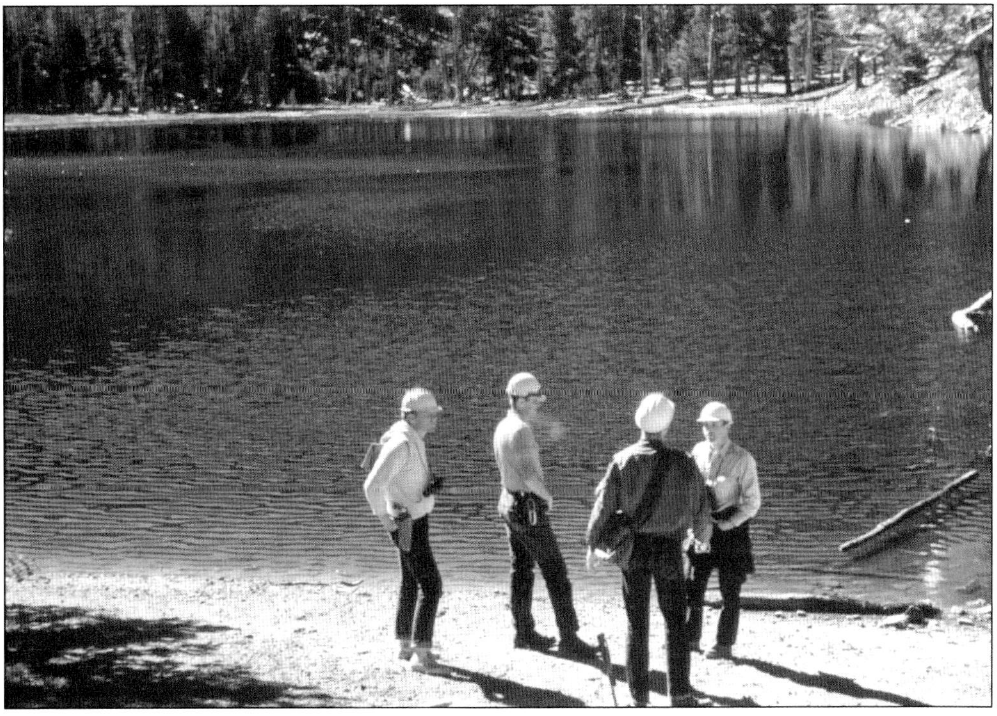

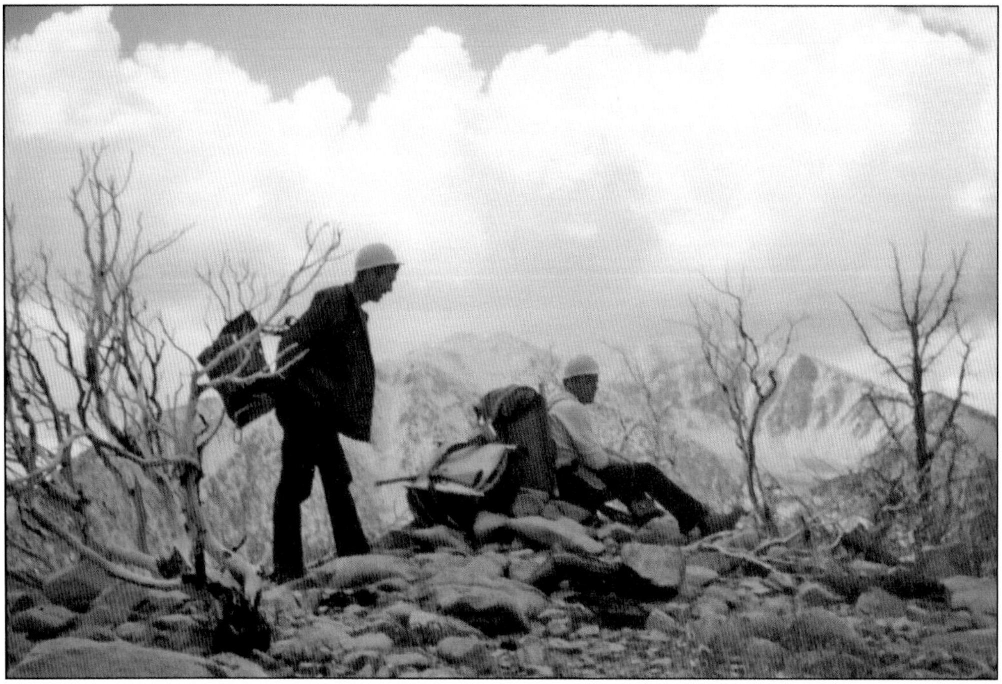

Searching the vast acreage that makes up the San Gorgonio Wilderness can be tedious and tiresome work. Team members take breaks frequently to look at maps, discuss where the lost hiker may be, and decide on a search strategy.

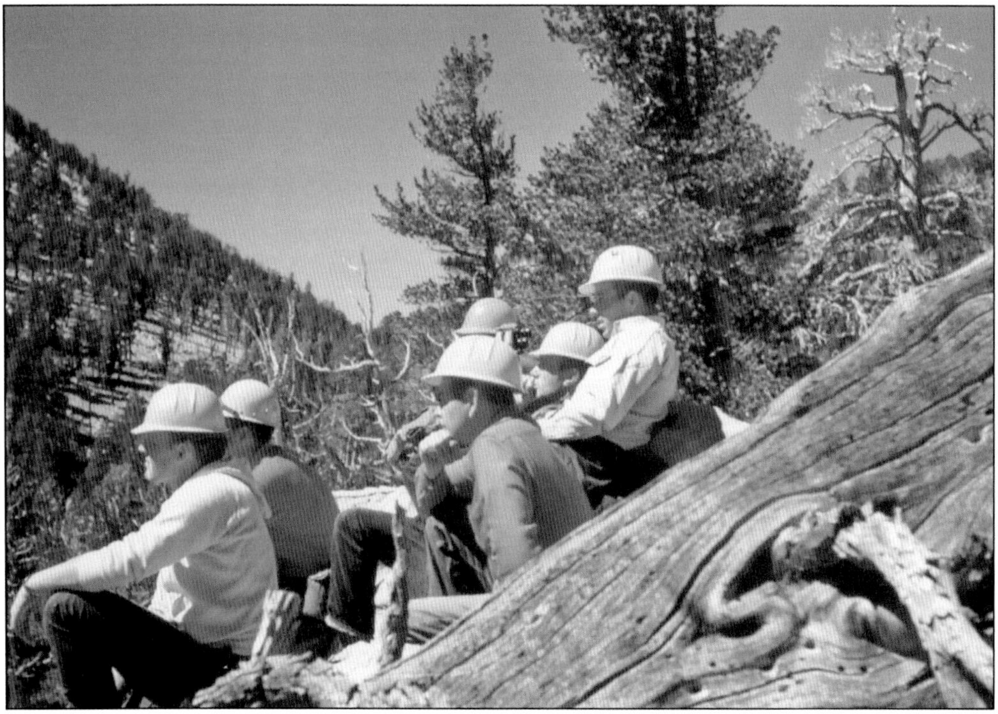

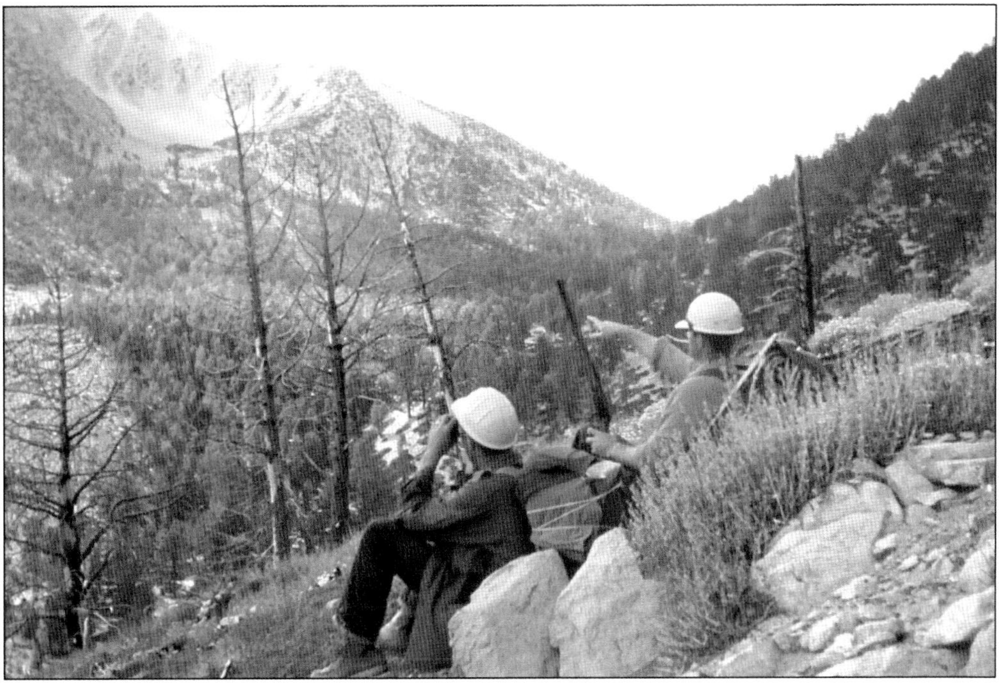

An added bonus during these searches, however, are the breathtaking views of the San Gorgonio Wilderness. The rugged mountains and gnarled trees help make the most arduous search a little easier for members of the search-and-rescue team.

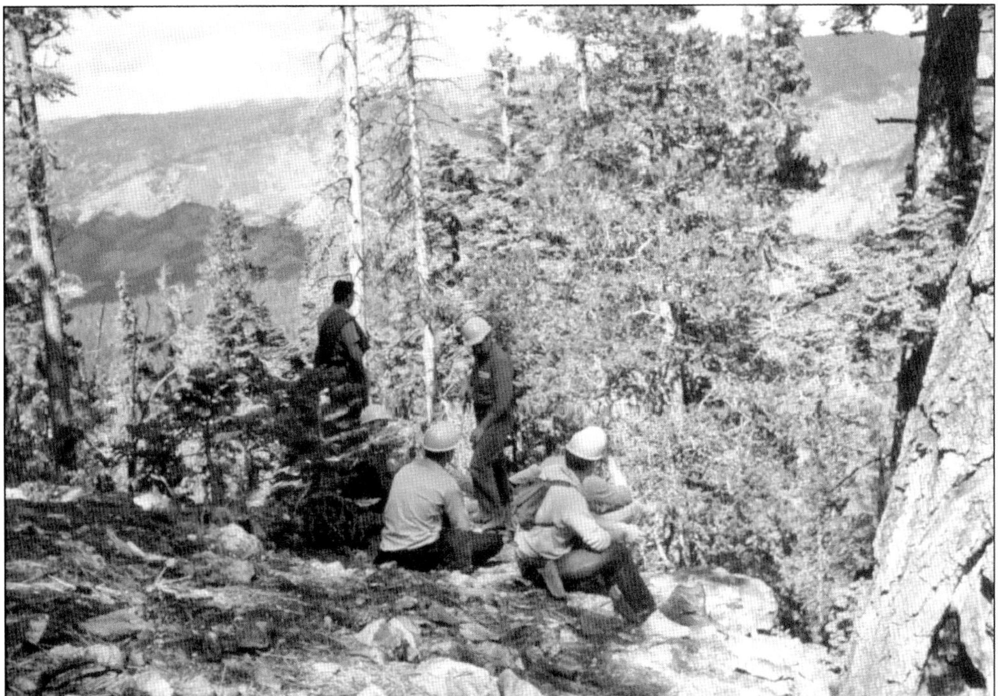

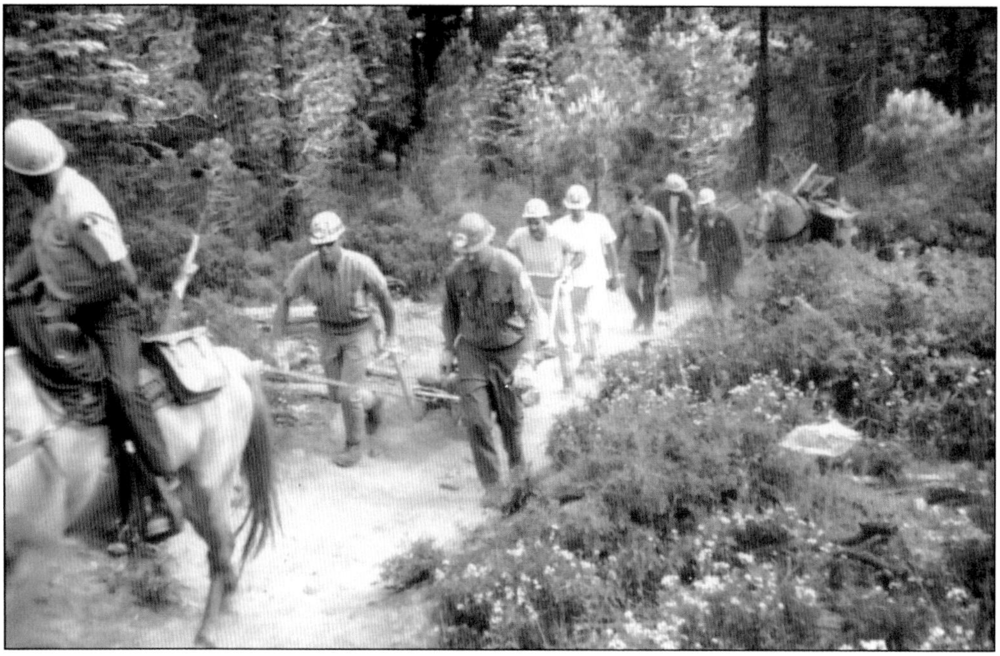

Here team members use horses and a wheeled litter to bring out an injured hiker on the South Fork Trail below Dry Lake, saving a lot of time and energy in getting injured people to medical attention. Today the team does not use horses. The sheriff's department has formed a mounted posse, which is its own volunteer unit similar to the search-and-rescue team.

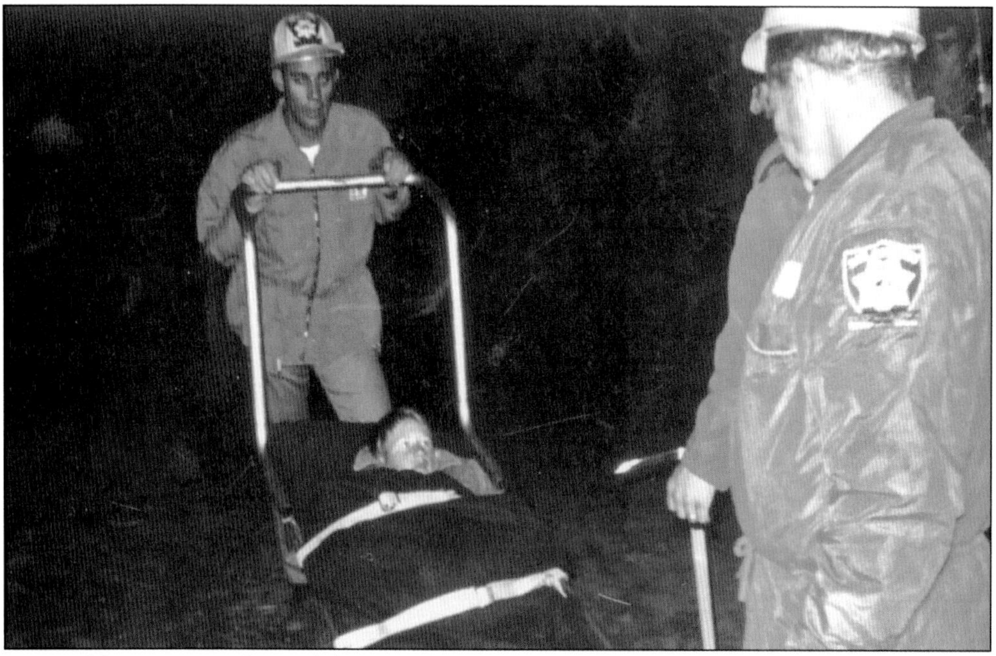

Pictured here are members of the team bringing an injured hiker out of the wilderness on a wheeled litter. The wheeled litter has been an integral part of the rescue team's equipment cache, enabling rescuers to transport injured hikers out of the wilderness more rapidly and with less physical exertion than a standard carry-out litter.

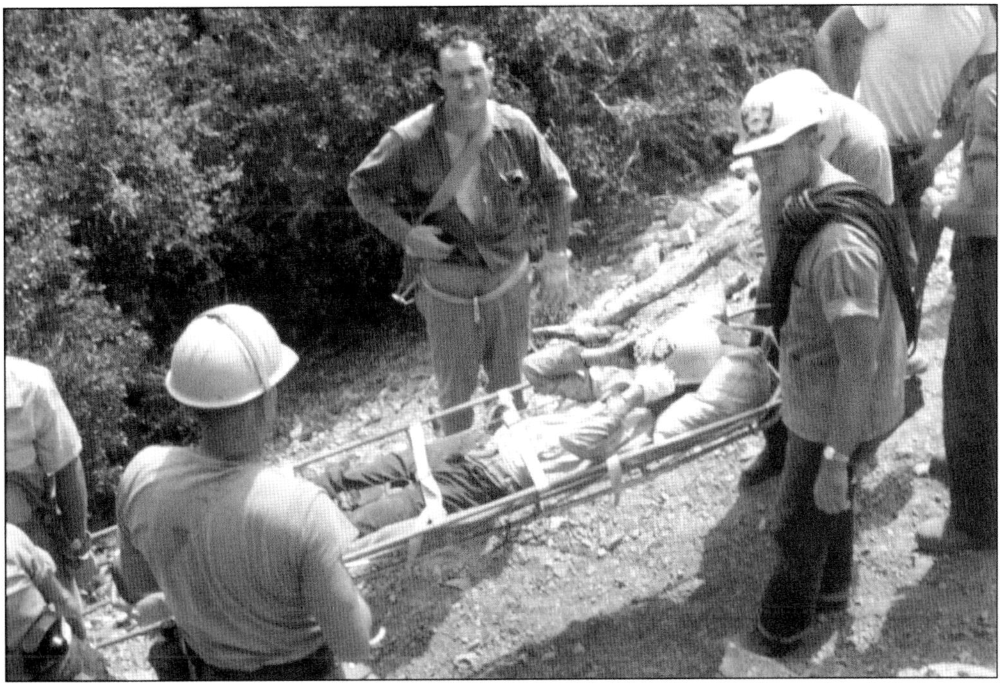

Here team members train by bringing an injured hiker up a scree slope near Vivian Creek Trail. Although the slope was not steep enough to warrant the use of ropes to haul the litter up the slope, the team did use a belay, or safety line, to keep the litter from falling back down the slope in case team members fell or dropped the litter.

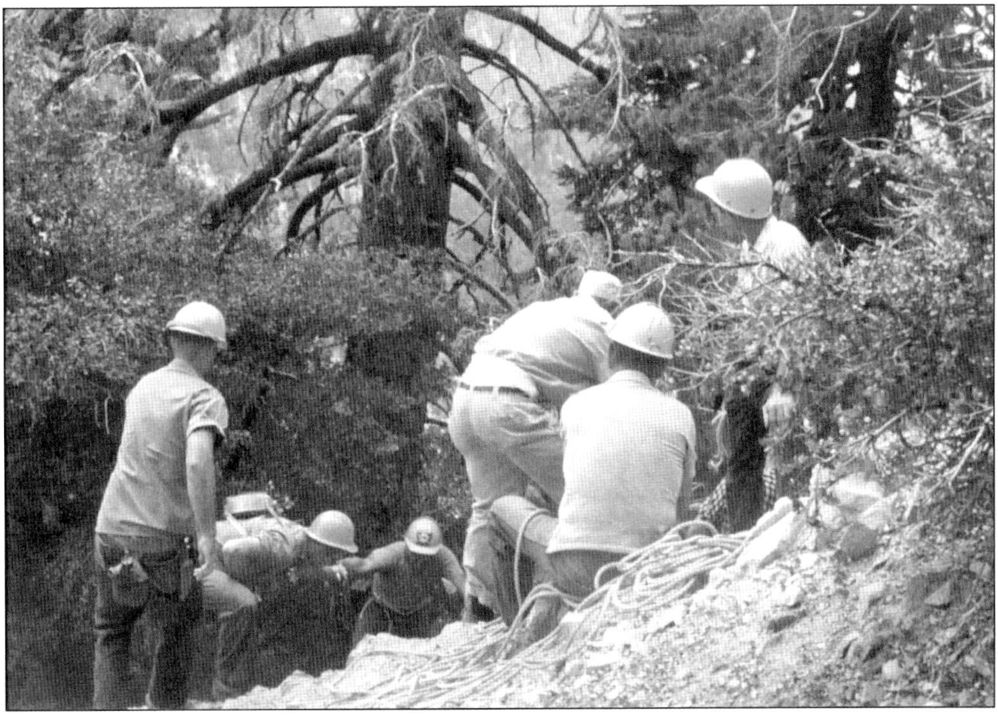

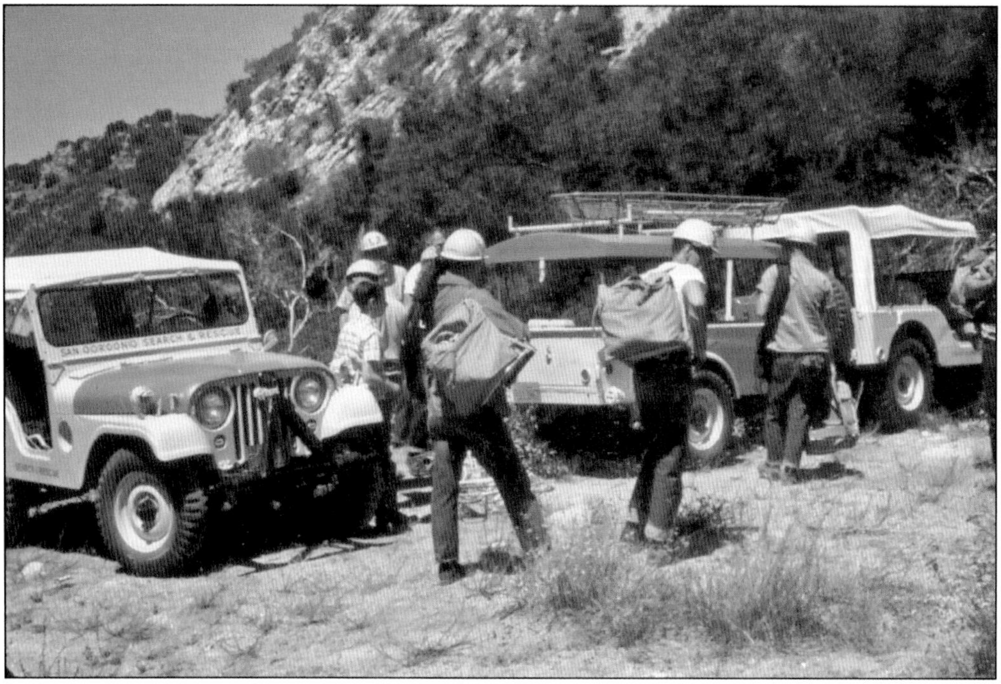

Before starting any search, team members will set up a staging area and a command post that will serve as the base of operations for the duration of the search. At the command post, the incident commander will organize searchers into teams and give them assignments for the areas they will search. Team members will secure their vehicles in the staging area and organize the gear they deem necessary given the type of search they are conducting and the possible condition of the person they are searching for.

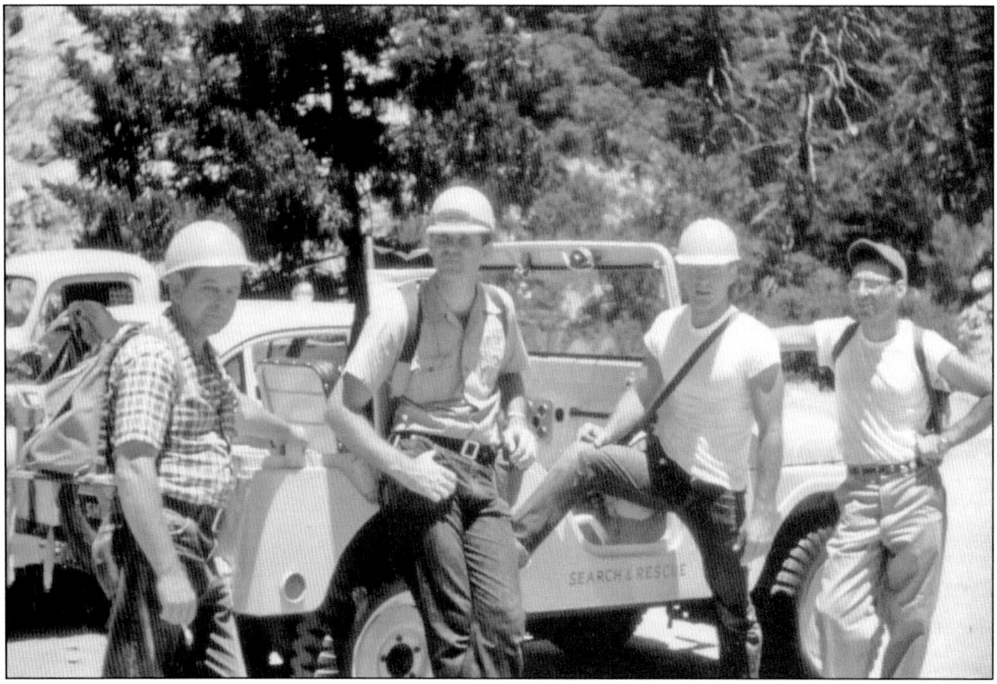

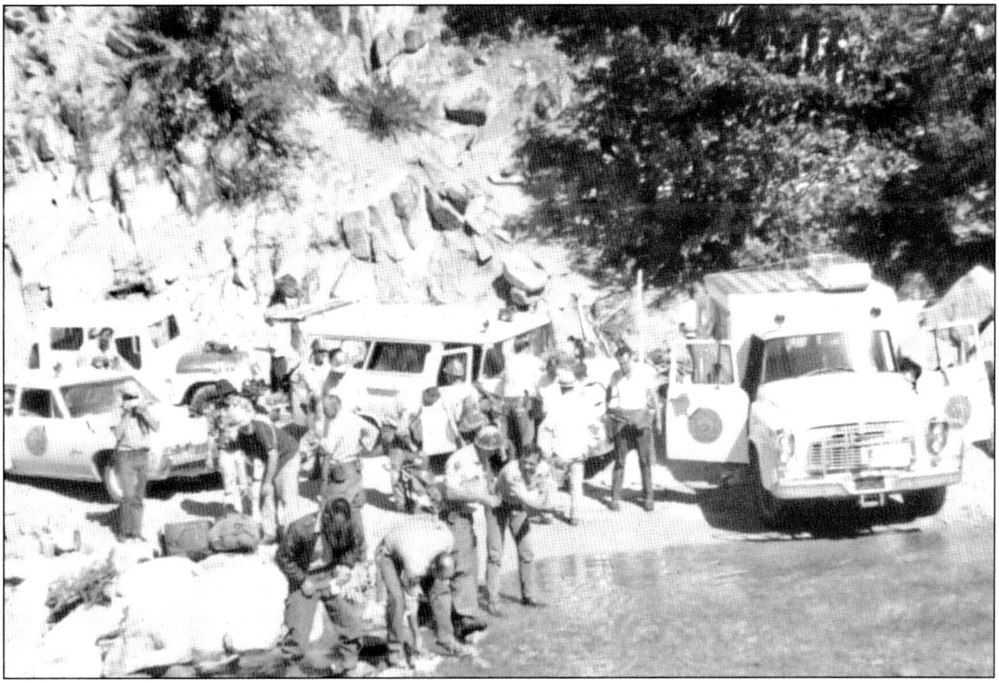

In the Mill Creek Wash, team members prepare their personal gear and equipment before heading into the wilderness to search for a lost hiker. Team members are carrying primitive canvas packs and filling their canteens out of Mill Creek. Note that the team still did not have a set uniform.

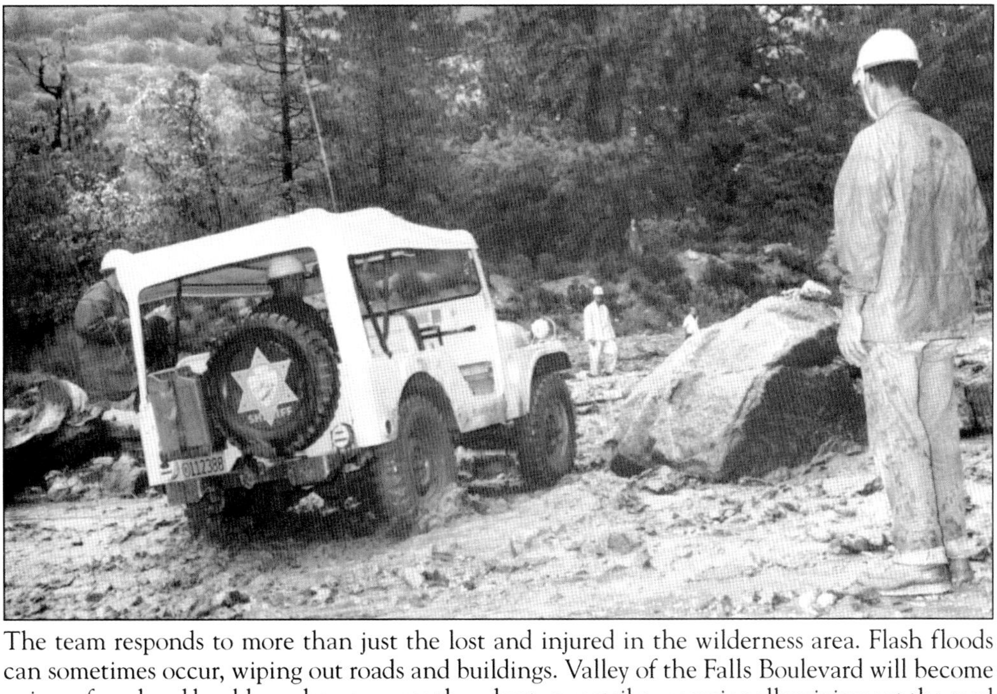

The team responds to more than just the lost and injured in the wilderness area. Flash floods can sometimes occur, wiping out roads and buildings. Valley of the Falls Boulevard will become a river of mud and boulders when summer thunderstorms strike, occasionally wiping out the road and taking with it the homes of canyon residents. The team's Jeep carefully negotiates a portion of the road in Mill Creek Canyon in 1962.

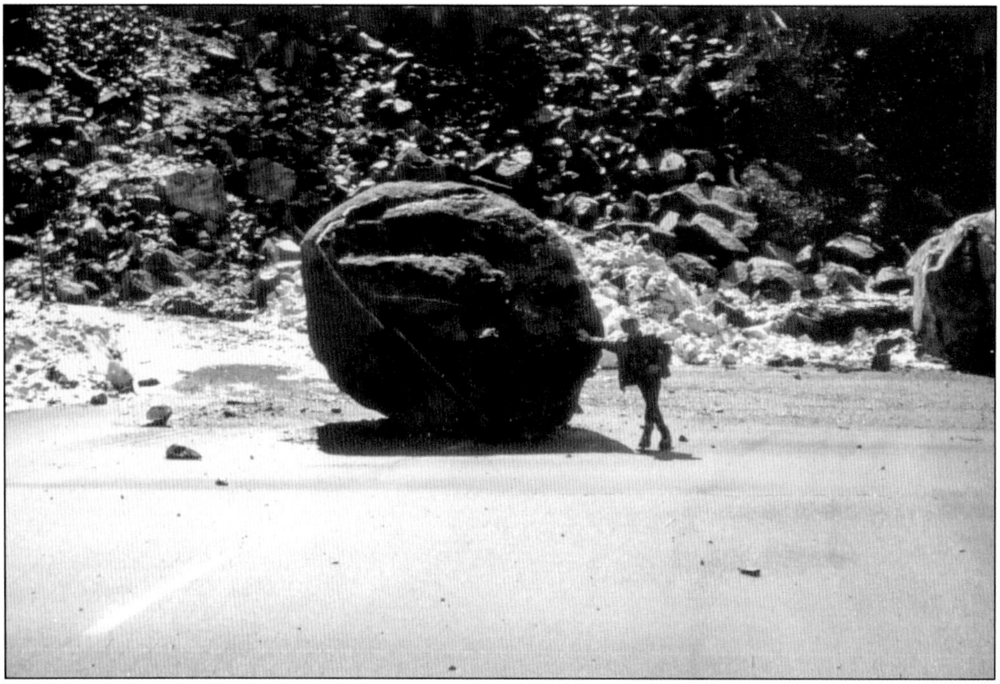

Portions of the San Andreas Fault run through the San Bernardino Mountains. The Landers–Big Bear earthquake of June 1992 was a 6.7-magnitude in Big Bear. A rock slide occurring in upper Mill Creek Canyon totally changed portions of Big Falls by falling rock. Some rocks were too large to be moved by equipment so they were eventually blasted into smaller pieces.

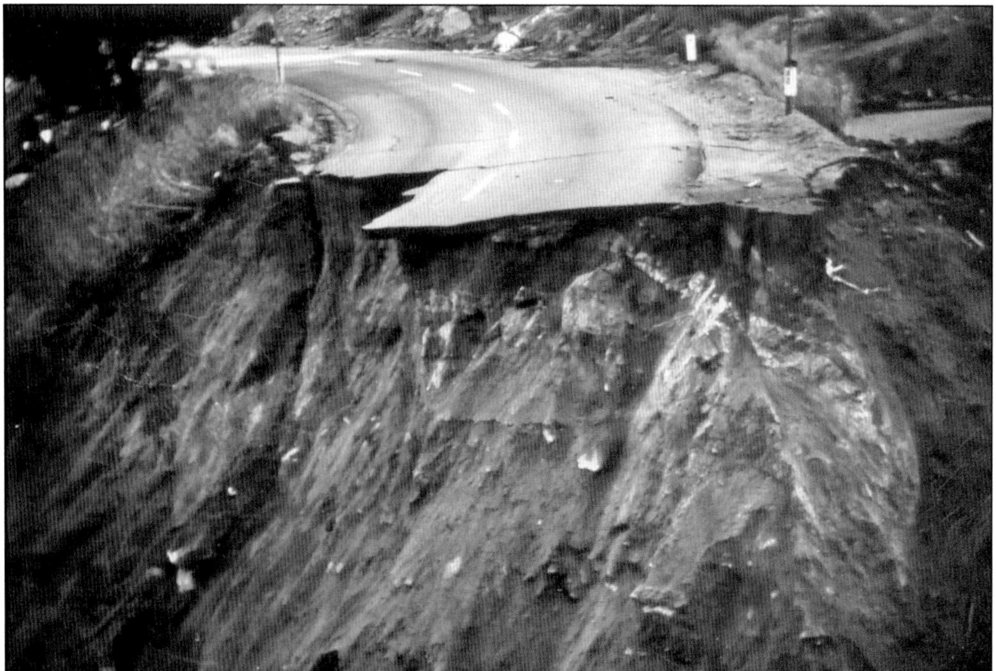

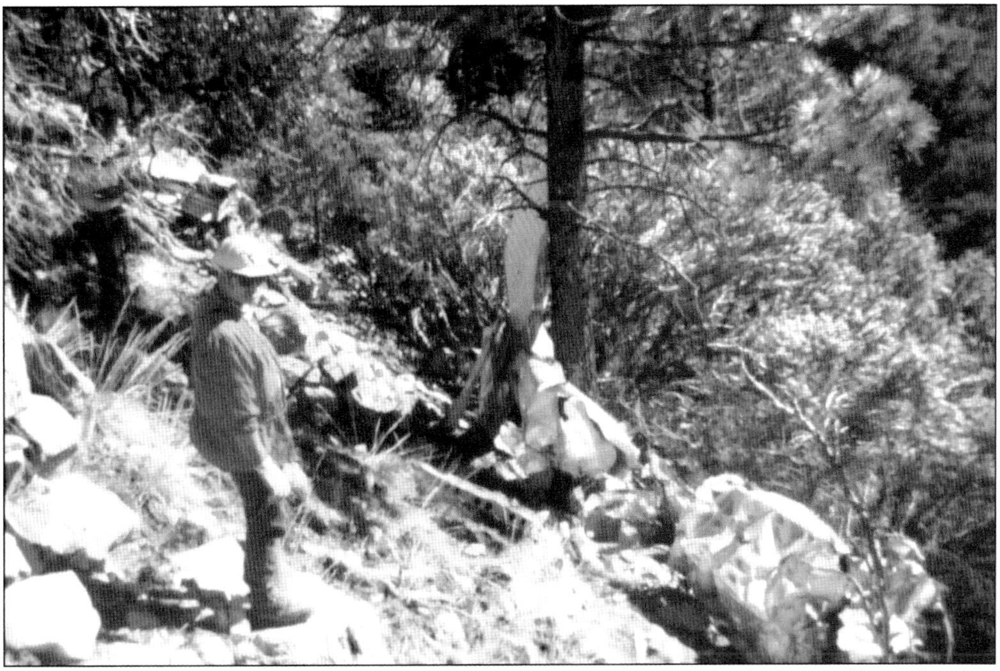

Plane crashes in the San Bernardino Mountains and surrounding foothills are all too familiar sights, and the crashes are often fatal. There have been some notable exceptions where victims walked away or were rescued alive from crash sites, but most of the time, the planes are hardly recognizable as aircraft, just twisted metal and Plexiglas.

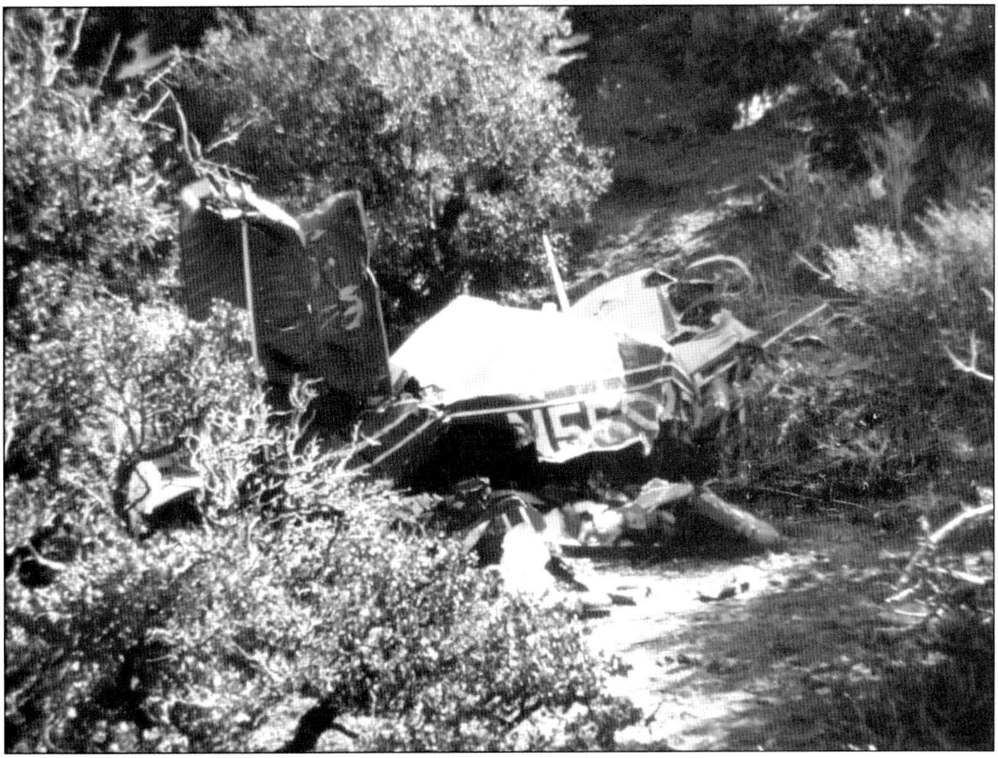

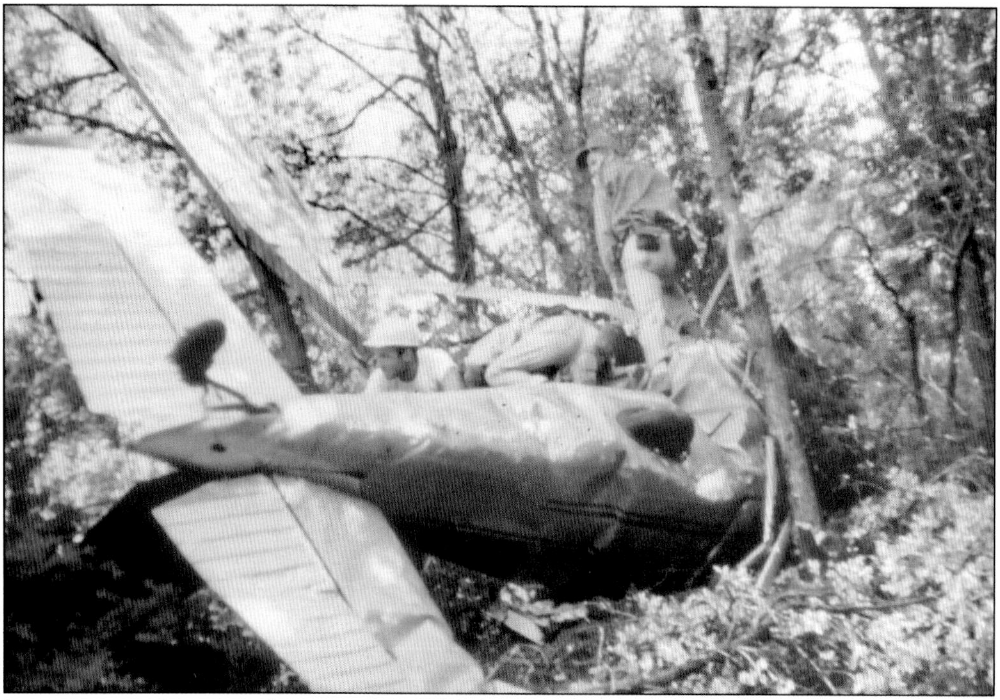

Sometimes crashes occur due to mechanical failure, but most are pilot error either from misjudging the terrain or weather conditions. Even experienced pilots have crashed in the wilderness. Some famous victims include Dolly Sinatra, mother of Frank Sinatra, whose Lear Jet crashed on the end of 10,000 Foot Ridge near Fish Creek Saddle, and Dean Martin Jr., whose Air National Guard jet crashed on the south facing slopes of the wilderness during a training flight.

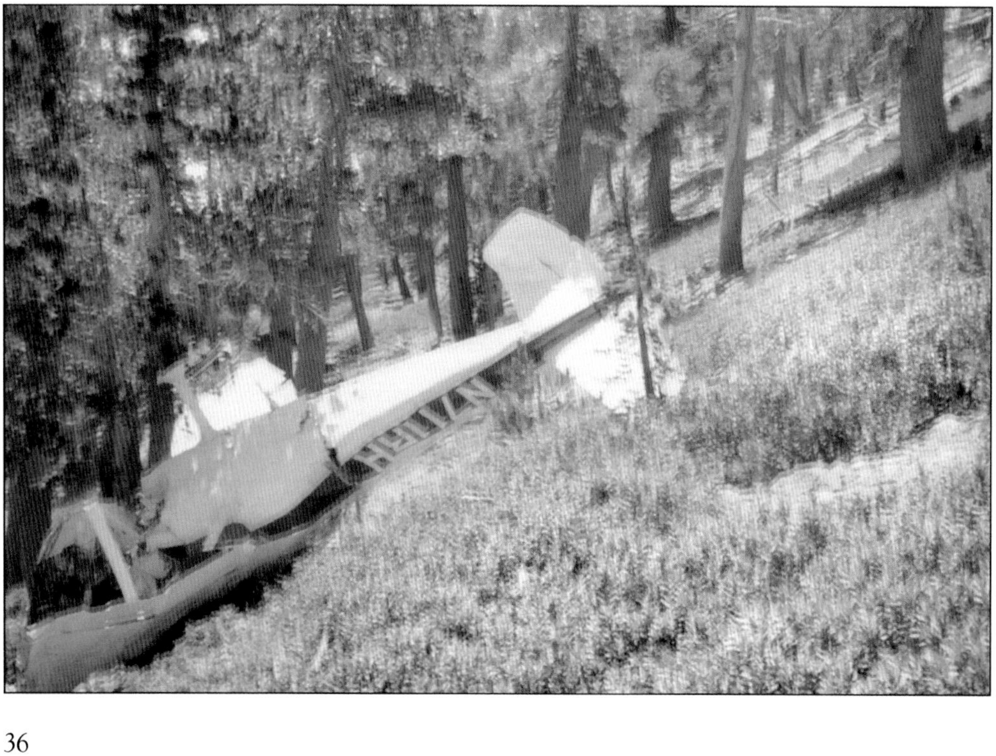

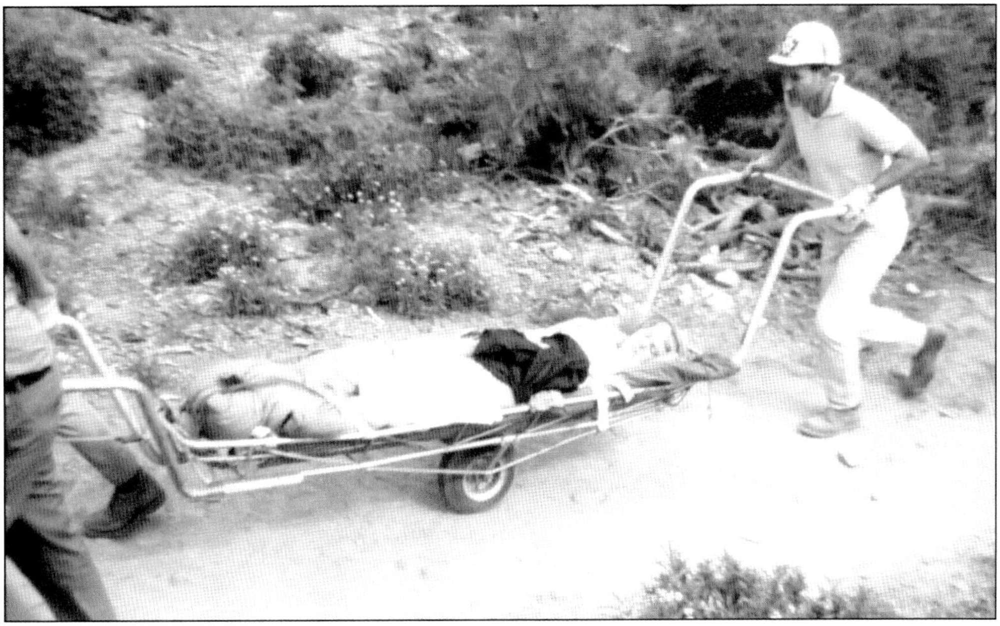

The team used a wheeled litter to quickly evacuate injured hikers from the wilderness, and this rescue in October 1967 was no exception. After locating the injured hiker, team members quickly got her medical attention. The wheeled litter was built from landing gear salvaged from a plane crash in the mountains.

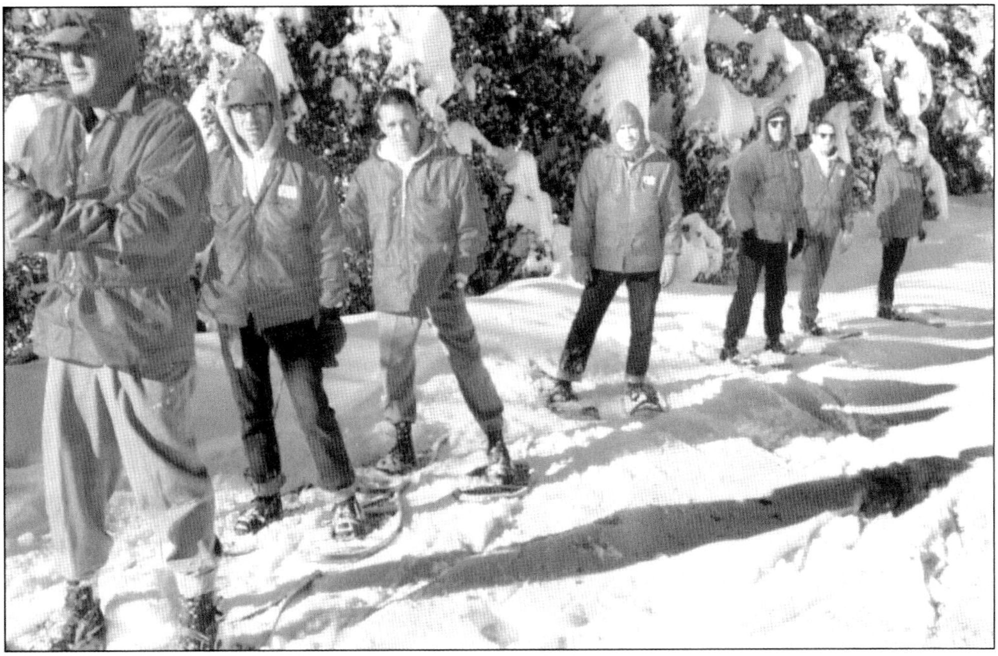

The San Gorgonio team trains together in the winter of 1961 near Barton Flats. There are two different types of snowshoes in use. The smaller ones with low front tips are an early version of the bear paw snowshoe. The longer ones with raised front toes are the old mountaineer snowshoes. Both styles are made of rawhide stretched between the wooden frames. This allows the snowshoe to float over soft snow and support the weight of the person.

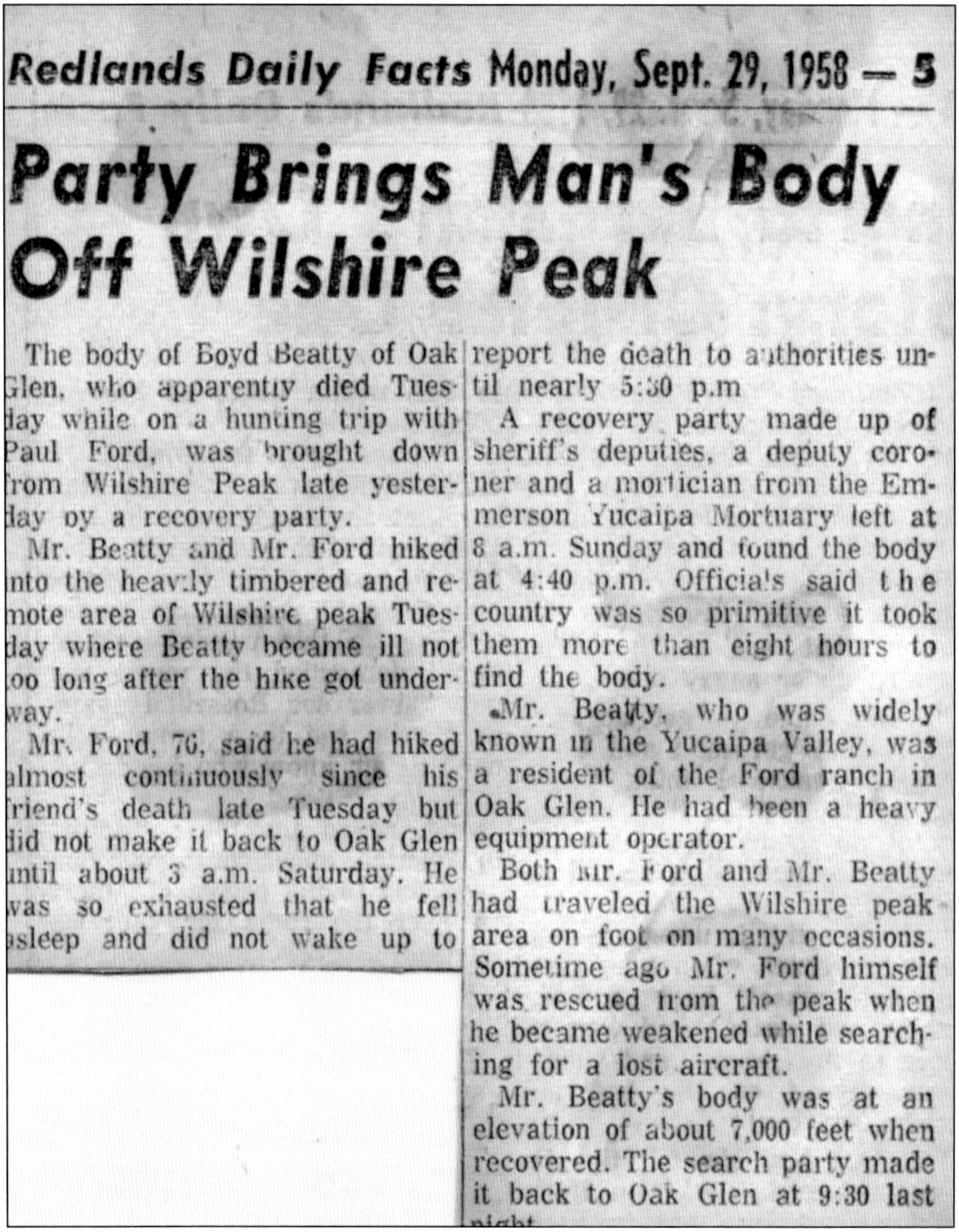

Redlands Daily Facts Monday, Sept. 29, 1958 — 5

Party Brings Man's Body Off Wilshire Peak

The body of Boyd Beatty of Oak Glen, who apparently died Tuesday while on a hunting trip with Paul Ford, was brought down from Wilshire Peak late yesterday by a recovery party.

Mr. Beatty and Mr. Ford hiked into the heavily timbered and remote area of Wilshire peak Tuesday where Beatty became ill not too long after the hike got underway.

Mr. Ford, 76, said he had hiked almost continuously since his friend's death late Tuesday but did not make it back to Oak Glen until about 3 a.m. Saturday. He was so exhausted that he fell asleep and did not wake up to report the death to authorities until nearly 5:30 p.m

A recovery party made up of sheriff's deputies, a deputy coroner and a mortician from the Emmerson Yucaipa Mortuary left at 8 a.m. Sunday and found the body at 4:40 p.m. Officials said the country was so primitive it took them more than eight hours to find the body.

Mr. Beatty, who was widely known in the Yucaipa Valley, was a resident of the Ford ranch in Oak Glen. He had been a heavy equipment operator.

Both Mr. Ford and Mr. Beatty had traveled the Wilshire peak area on foot on many occasions. Sometime ago Mr. Ford himself was rescued from the peak when he became weakened while searching for a lost aircraft.

Mr. Beatty's body was at an elevation of about 7,000 feet when recovered. The search party made it back to Oak Glen at 9:30 last night.

Recovery of the dead is part of the duties for the rescue team. In this case, the victim was on a hunting trip in a remote and rugged area of the San Bernardino Mountains on the Yucaipa Ridge. The area was so primitive that it took the team members more than eight hours of hiking to locate the body. Once located, it took the team over five hours to bring him out to Oak Glen.

Three

Team Equipment over the Years

The San Bernardino County Board of Supervisors approved Sheriff Frank Bland's proposal to form the county's first mountain search-and-rescue team, but they only approved limited funding. Due to the sporadic and inconsistent nature of the job, team members would be volunteers who could be called out at any time to leave their jobs and families to search for the lost and injured in the mountains. The team would be responsible for purchasing all of their own equipment and vehicles and finding their own base of operations to hold team meetings and house their vehicles and equipment. The county board approved funding for maintaining the team vehicles and replacing equipment on a limited basis. From its first months, the team has conducted public relations events to raise the funds needed to support the team. Team members also purchase most of their personal equipment with money out of their own pockets.

Through the years, the rescue equipment that the team has used has evolved. In the beginning, the team used basic hiking and camping gear and military surplus rope gear. As the public interest in outdoor recreation grew, the outdoor equipment industry has grown as well. Companies developed lighter, warmer, and more durable outdoor equipment. As outdoor recreation has become more popular and more accessible to the general public, more people have found themselves in situations that require the aid of search-and-rescue teams like San Gorgonio Search and Rescue.

As the need worldwide has grown for quality rescue equipment, the outdoor industry has risen to meet the need by developing a whole line of outdoor equipment specifically designed for the search-and-rescue community. Many rock climbing gear companies have also taken on the development and production of rope rescue equipment and make rock climbing equipment that is specifically designed for mountain search-and-rescue teams. As the outdoor recreation industry has changed, so has the rescue equipment that the San Gorgonio Search and Rescue Team uses.

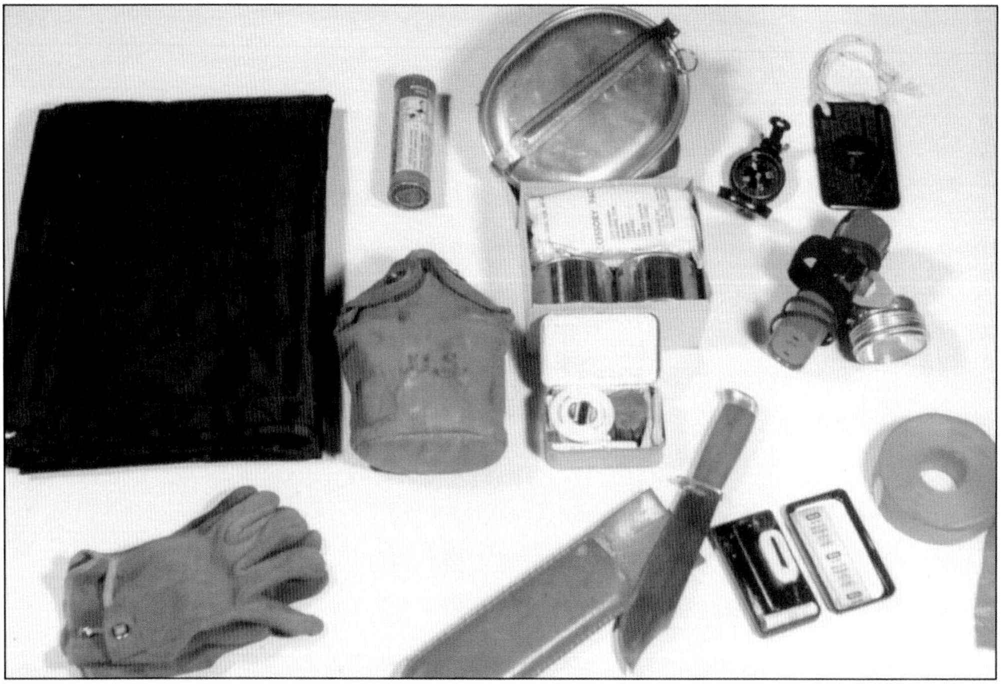

Pictured here is an example of the equipment some of the original team members carried in their backpacks. From top left are a tarp, gloves, a flare, canteen, mess kit, food rations, first aid kit, knife, compass, reflective mirror, flashlight, and flagging tape.

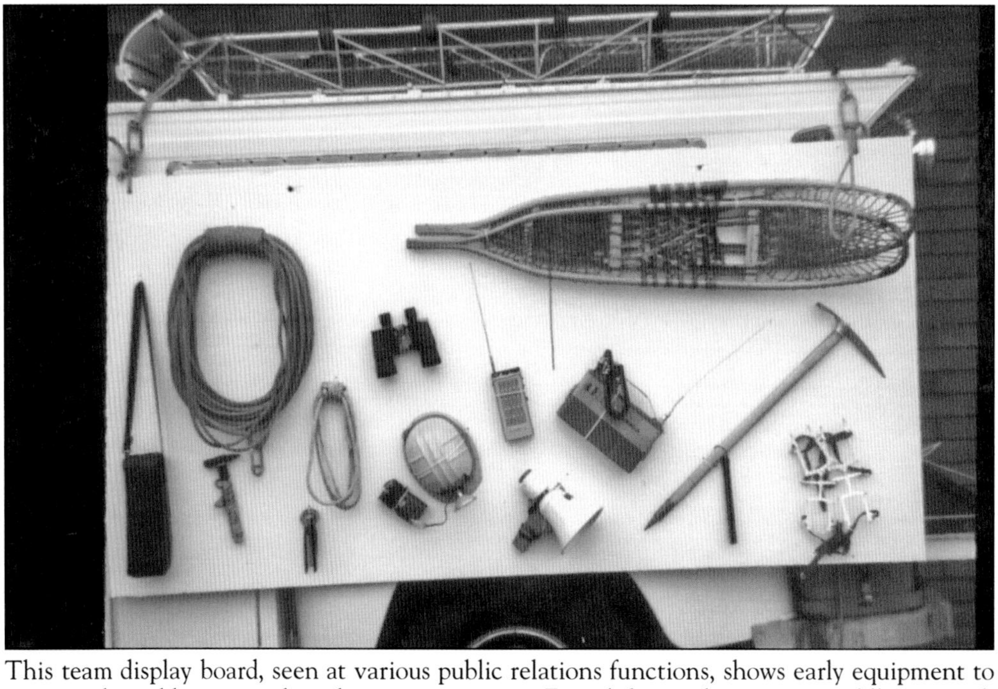

This team display board, seen at various public relations functions, shows early equipment to instruct the public in search-and-rescue operations. From left to right are army goldline, a rock hammer, rope runners, binoculars, pitons, snowshoes, an avalanche transceiver, a radio, a bullhorn, an ice ax, and crampons.

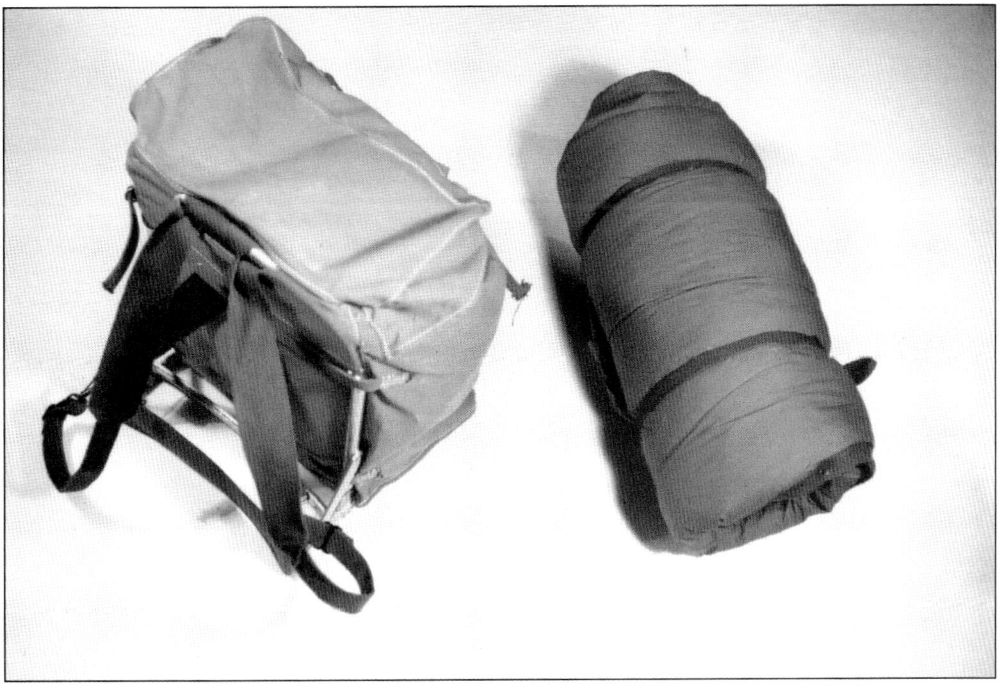

The first backpacks worn by the team had a metal frame with a canvas bag attached to it. Canvas shoulder straps put the weight of the pack on the back, but there was no waist belt to place the weight on the hips like today's backpacks. Team members carried a small cotton sleeping bag filled with goose down, not very warm compared to today's synthetic and nylon sleeping bags. They also carried a canteen, a first aid kit, an ax, and a large bulky radio.

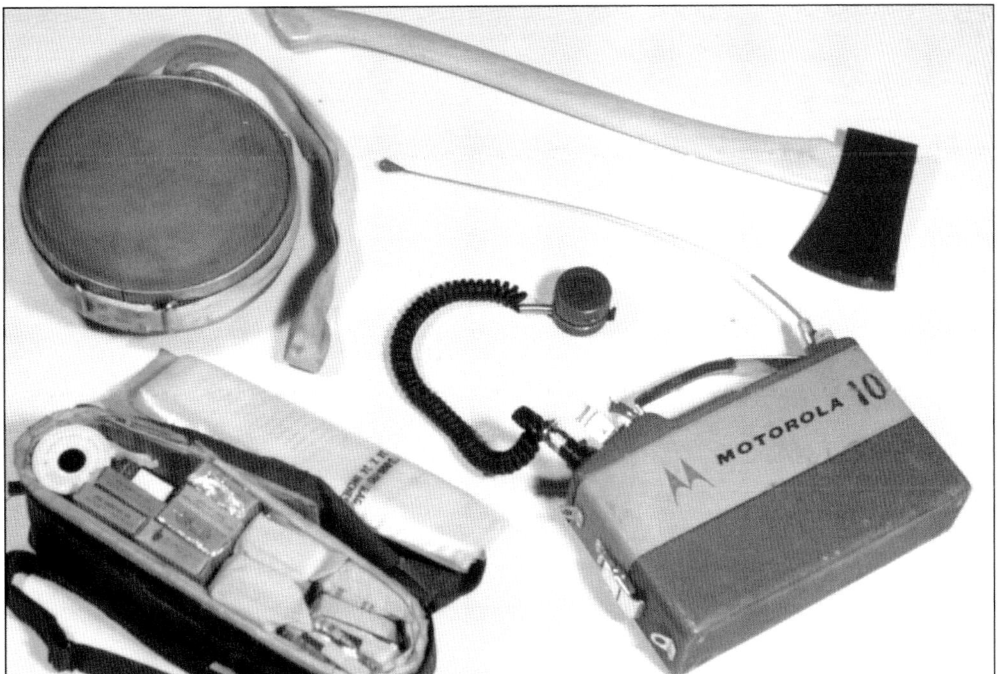

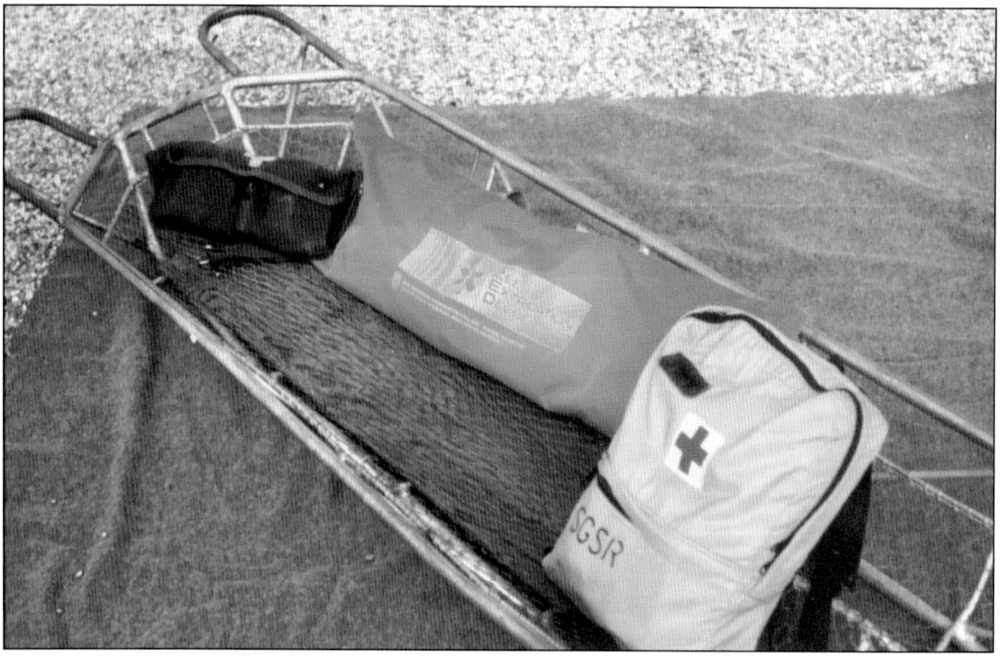

Pictured here are early litters and first aid kits used by the team to rescue victims who had fallen off steep cliffs. The original litters were made of steel and weighed more than 30 pounds; today they are made of titanium and weigh about eight pounds. They also break into two pieces so they can be carried more easily into the wilderness.

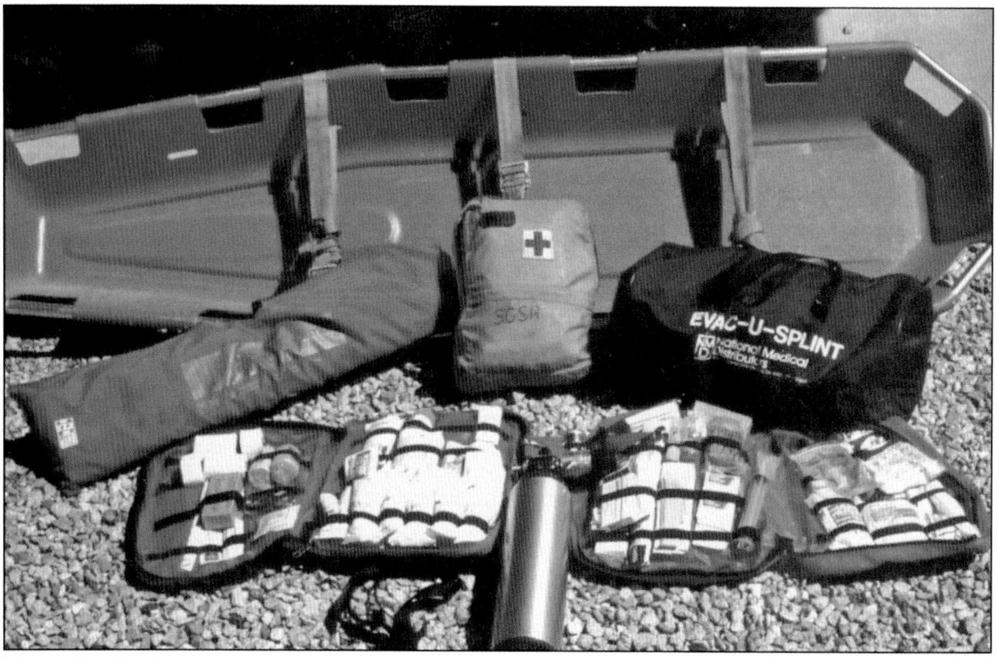

First aid technology continues to improve, and as it does, the team upgrades their first aid equipment to better meet the needs of those who are injured in the wilderness. Compared to the early days, the team uses lightweight plastic stretchers and litters to carry the injured out of the wilderness as well as lightweight and compact first aid kits.

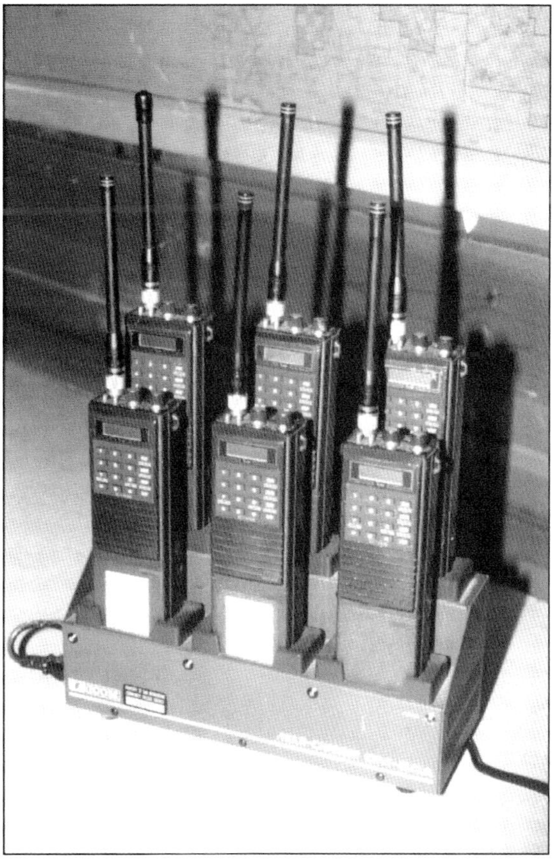

The team uses a combination of 155mhz and 800mhz radios to communicate with one another and other agencies while engaged in rescue operations. Radios are mounted in the mobile command post (a converted bread truck), and team members carry handheld radios into the field.

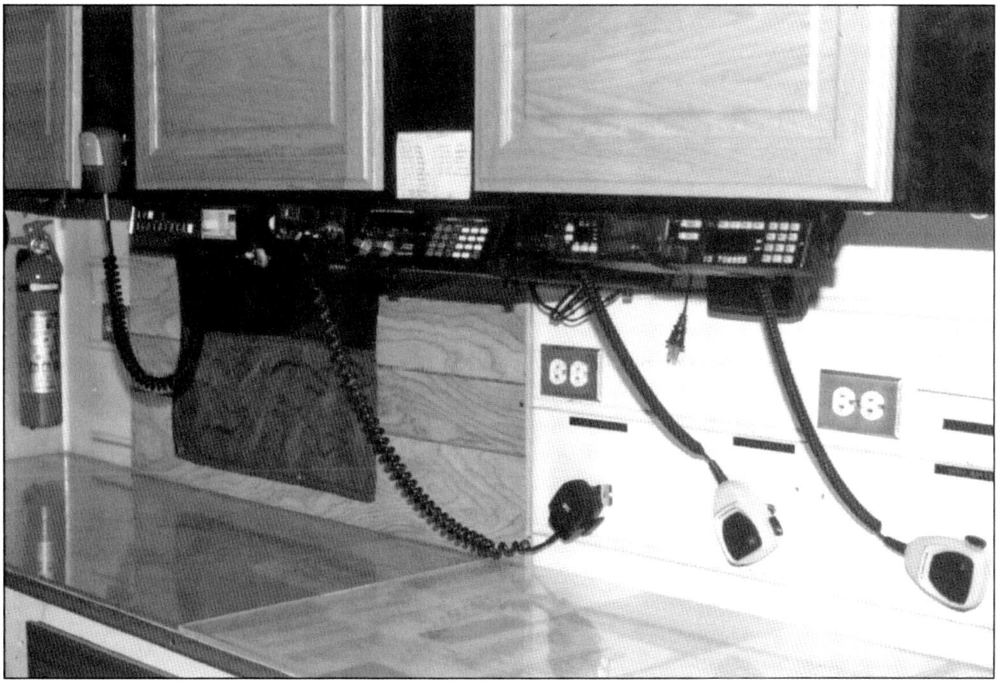

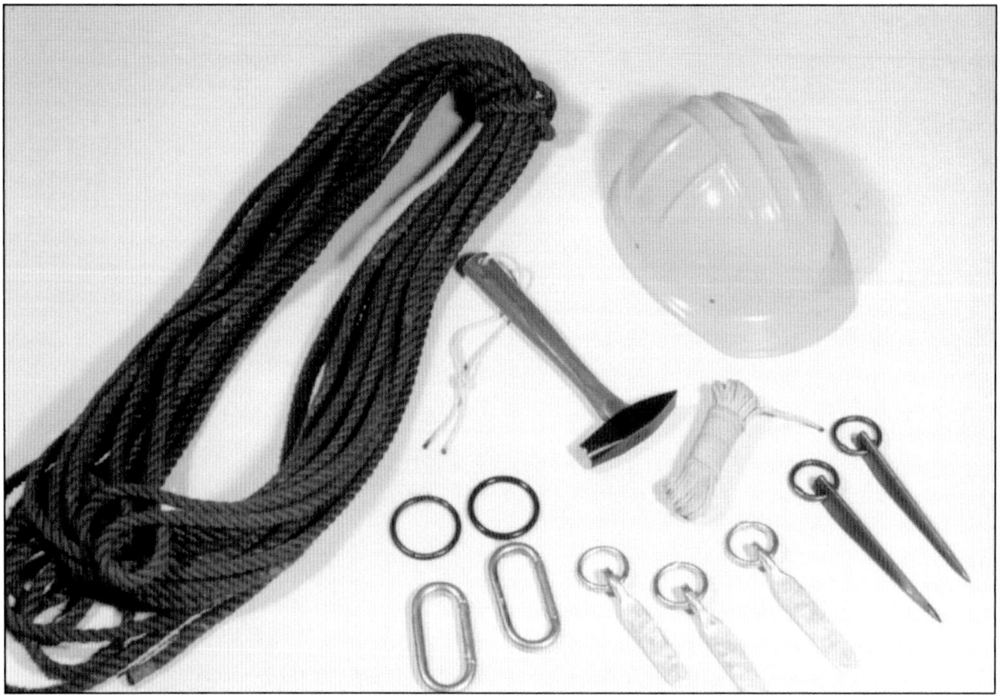

The yellow construction helmet was a team standard for many years. Later the construction helmets were replaced with the orange John Brown helmets. The team also began using more sophisticated pulleys, figure eight rappel devices, and woven seat harnesses.

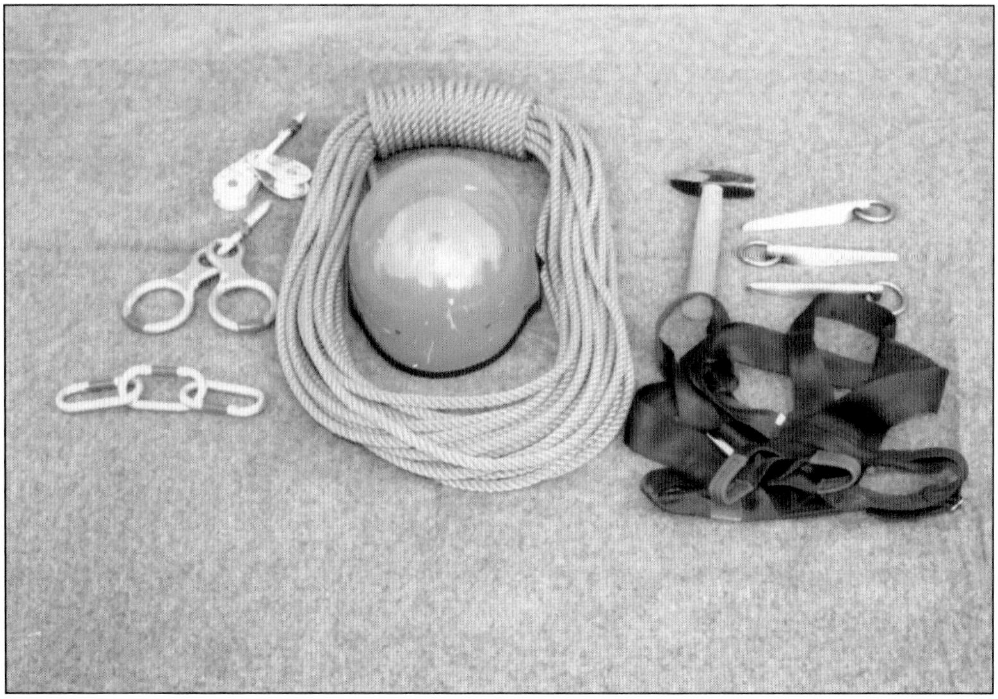

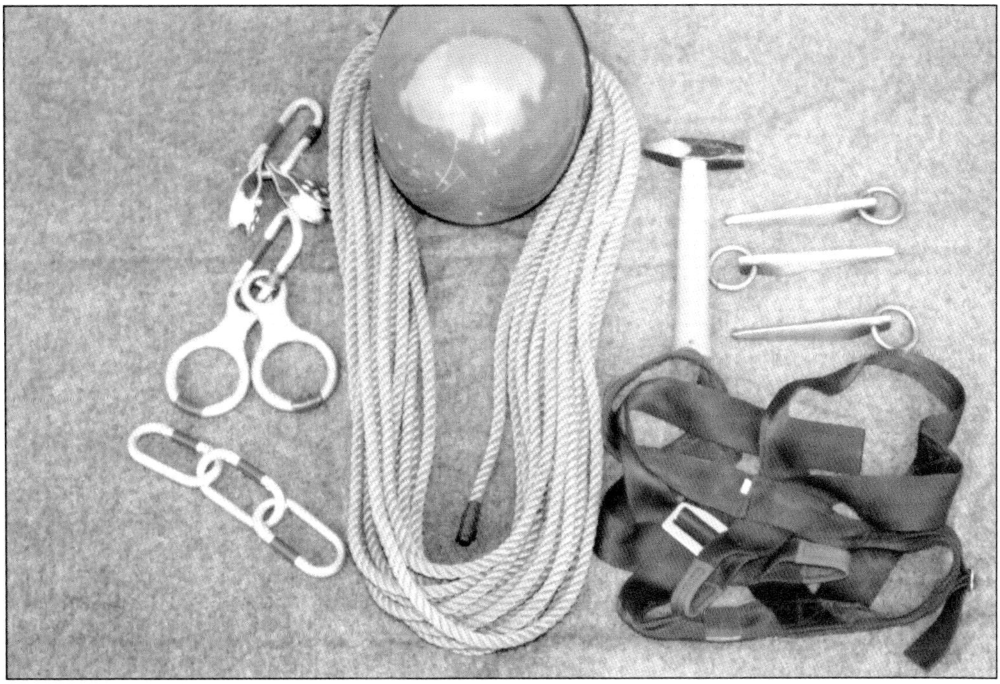

Along with army goldline, the team also used pitons along with primitive carabineers and pulleys to set up anchors on the tops of steep cliffs to perform rope-rescue operations.

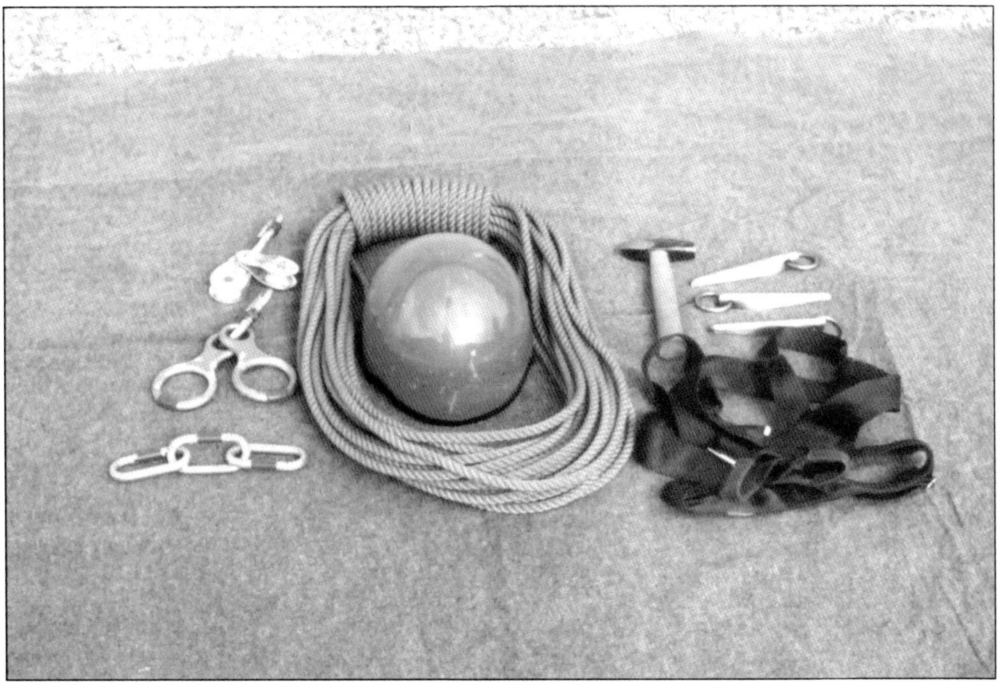

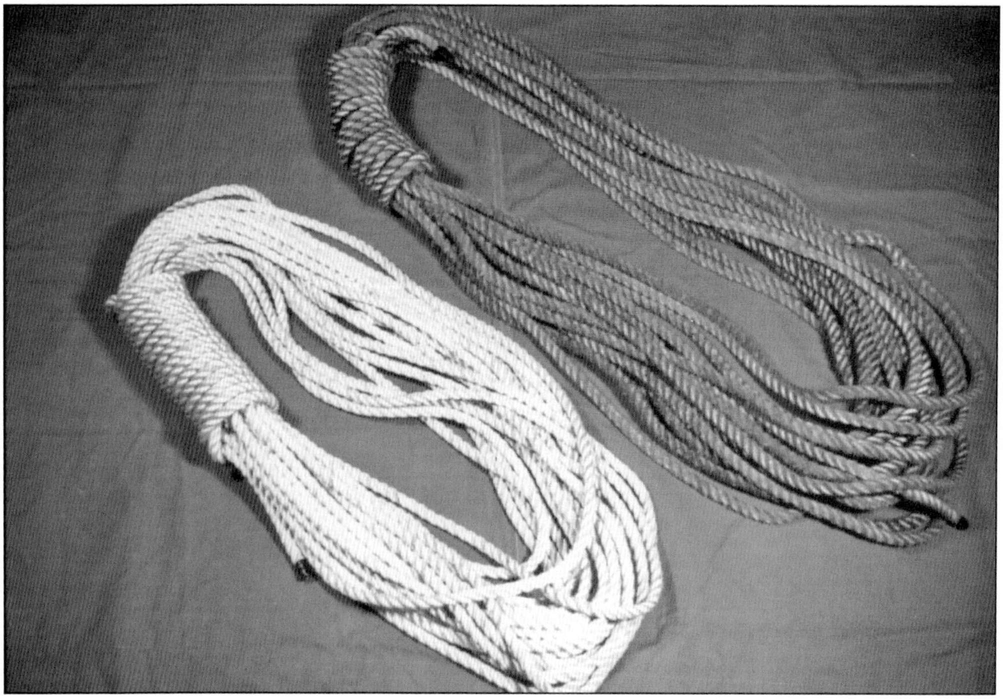

In the early years, the team used army surplus goldline and static nylon rope for all of its rope-rescue trainings and operations. Rope-rescue technology has come a long way since the 1950s, and the team now uses 11-mm and 12-mm kern-mantel static rescue rope. Kern mantel is the standard rope used for rescues and rock climbing. It is made from nylon strands woven together and covered by a nylon sheath.

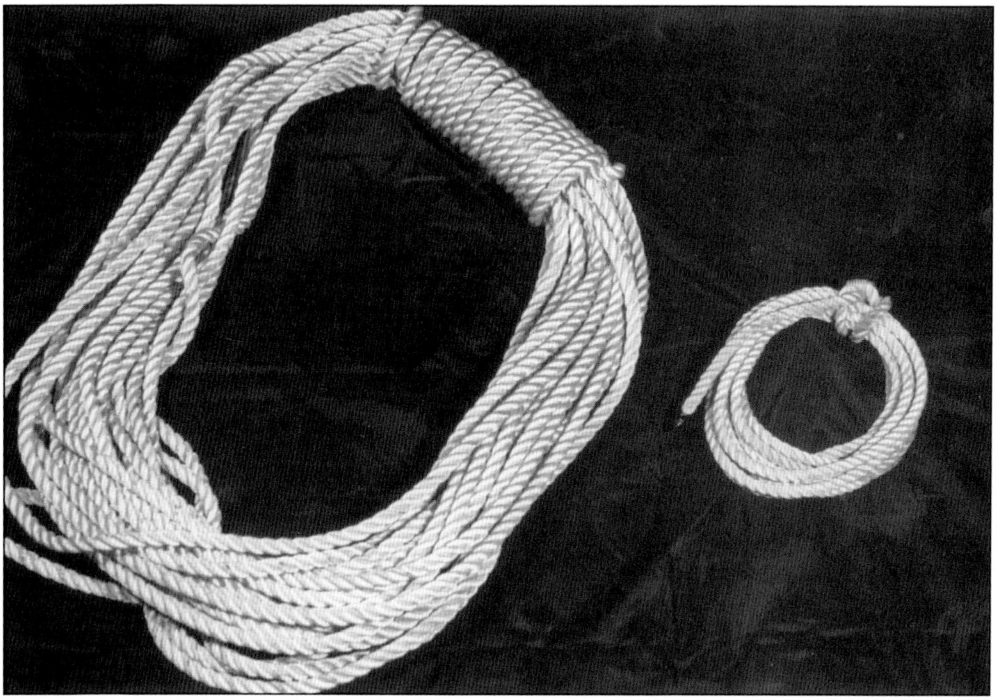

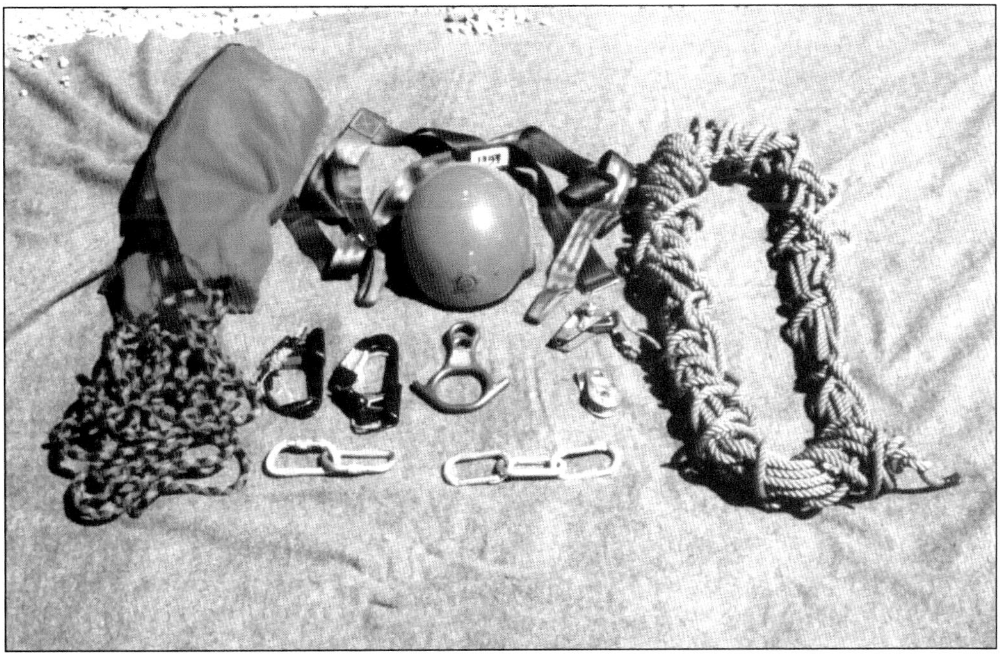

In later years, as rope-rescue technology improved, the team began using static 9-mm prusik cord for a variety of rope-rescue situations—ascenders for climbing the rope, locking carabiners, rescue figure eight rappel devices, and woven seat harnesses. The team continues to update its rope-rescue equipment and uses the latest equipment in rescue technology.

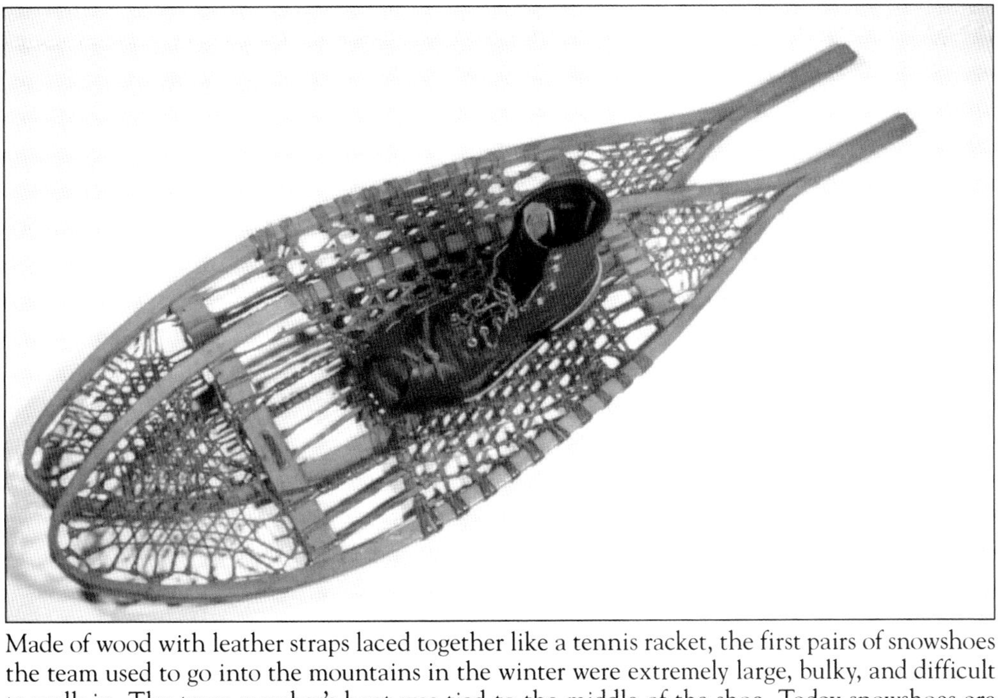

Made of wood with leather straps laced together like a tennis racket, the first pairs of snowshoes the team used to go into the mountains in the winter were extremely large, bulky, and difficult to walk in. The team member's boot was tied to the middle of the shoe. Today snowshoes are made of nylon cloth stretched around an aluminum alloy frame. They are much lighter, and the boots strap in with secure bindings.

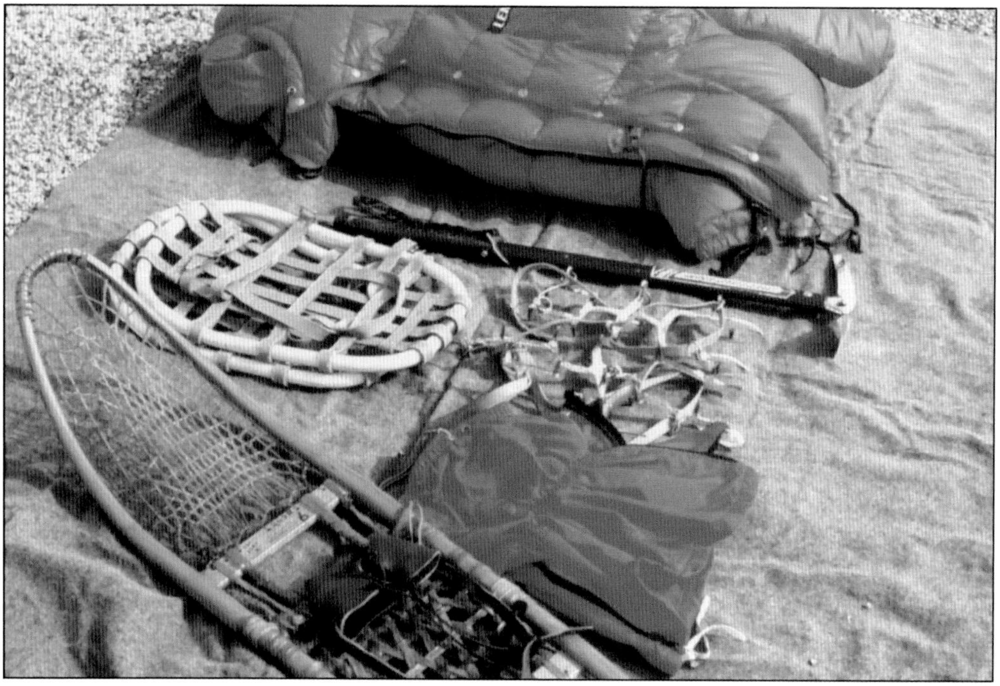

Ice axes and crampons were standard team equipment for searching for lost and injured hikers in the snow. The original ice axes were made out of wood and forged steel, and crampons were tied to the bottom of team members' boots. Today's ice axes are made from lightweight aluminum alloys, and crampons are custom fitted to mountaineering boots.

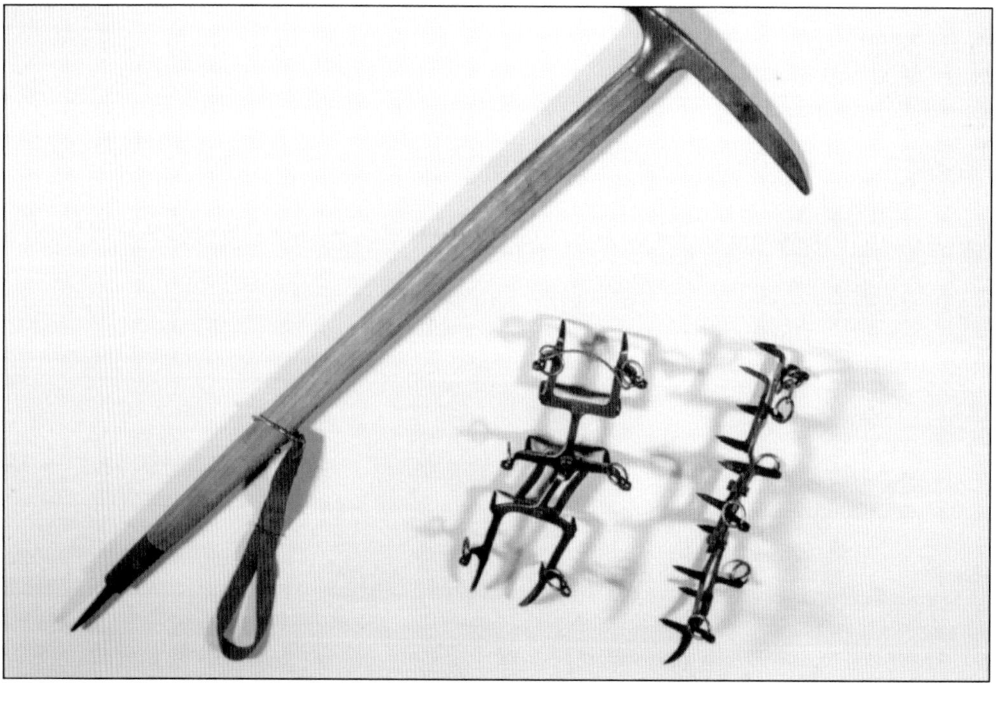

Four

A History of Rope Rescue

The San Gorgonio Search and Rescue Team was founded because of the need for qualified rope-rescue personnel in San Bernardino County. Therefore rope rescue has always been a major priority for the team. When the team began, there were not national, state, or county standards that governed rope-rescue operations. There were no organizations that taught rope-rescue techniques and no real established guidelines for rope-rescue procedures. Team members relied on personal experience from the military and recreational rock climbing and mountaineering to guide them as they trained for situations like the one that killed Donald Burns on Big Falls.

For many years, the team trained and certified its own members in rope-rescue techniques. As San Bernardino County expanded its volunteer search-and-rescue program, other teams were formed, such as Valley of the Falls, Big Bear, Central, and West Valley. These teams were also left to their own expertise when it came to rope-rescue training. It was not until the late 1990s that San Bernardino County began to standardize its rope-rescue program and develop county-wide procedures for all rescue teams. The county developed a Rope Rescue Oversight Committee (RROC) made up of volunteers from various search-and-rescue teams. This committee turned to other counties with successful rope-rescue teams as well as to nationally recognized rope-rescue training programs developed by the military and rescue companies. The RROC attended classes by nationally recognized rope-rescue schools and brought what they learned back to the county to develop a county-wide rope-rescue training program. In 1997, members of the team attended the first Technical Ropes Basic Course put on the by the county. The county also developed rope-rescue standards and a test each team needed to pass in order to be certified by the county to perform rope-rescue operations. The team passed their rope-rescue certification in October 1998 and has successfully retested every two years since then to maintain their certification status.

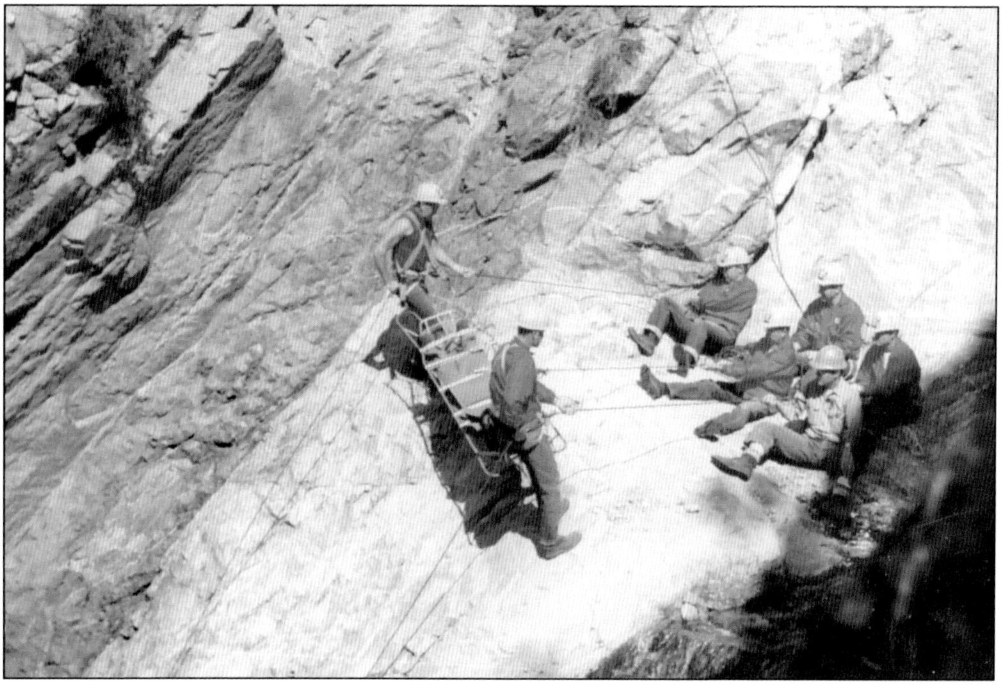

At the top of a steep cliff near Monkey Face Falls in Mill Creek Canyon team members set up ropes to train on the almost vertical rock face. Team members practiced rappelling and lowering and raising a litter to prepare for the inevitable call that another climber has injured himself on the falls.

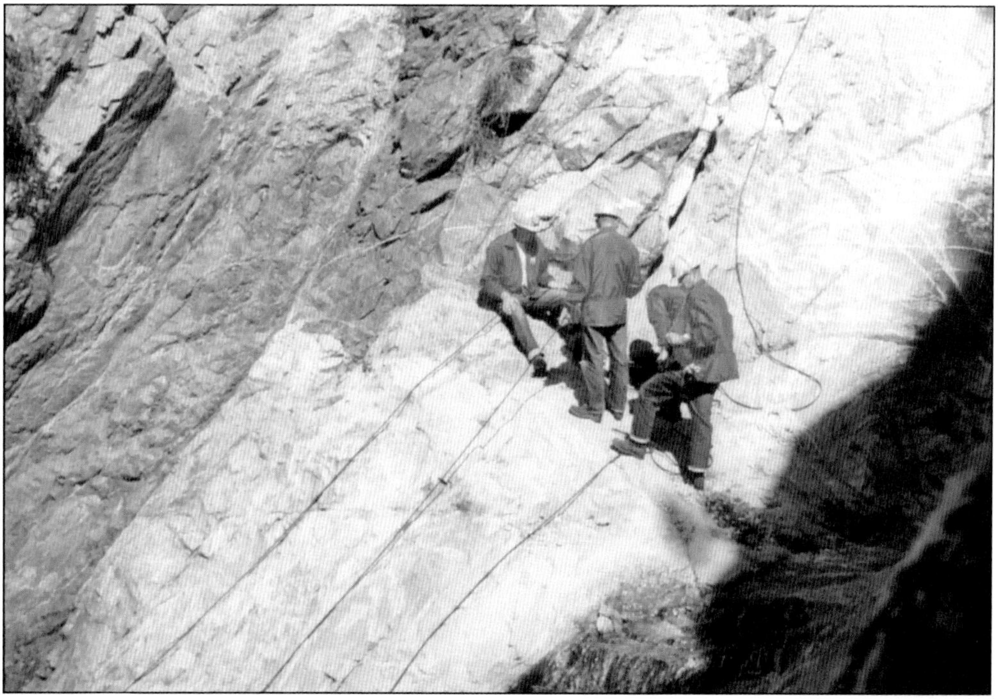

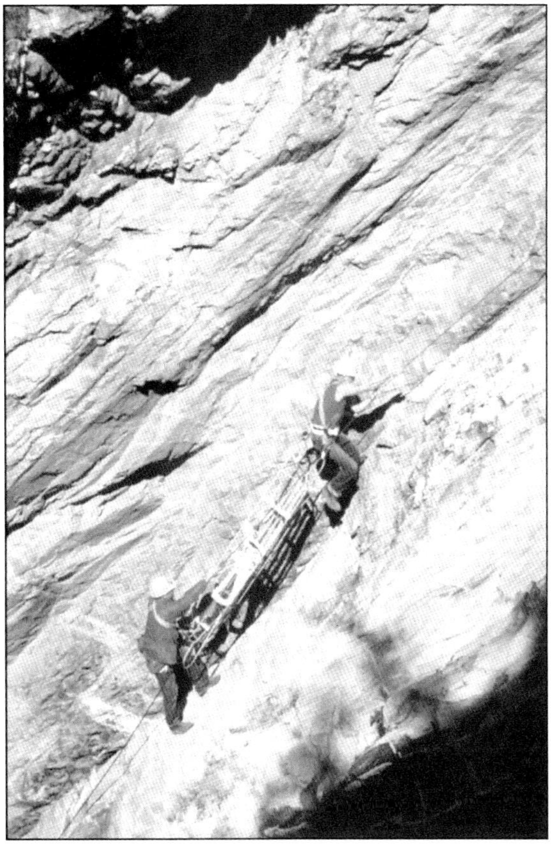

Being lowered down one of the sheer rock faces near Big Falls in Mill Creek Canyon, rescue team members train repeatedly until the rope-rescue procedures are second nature to them; in a real rescue in a vertical environment rescuers do not have the option of making a mistake.

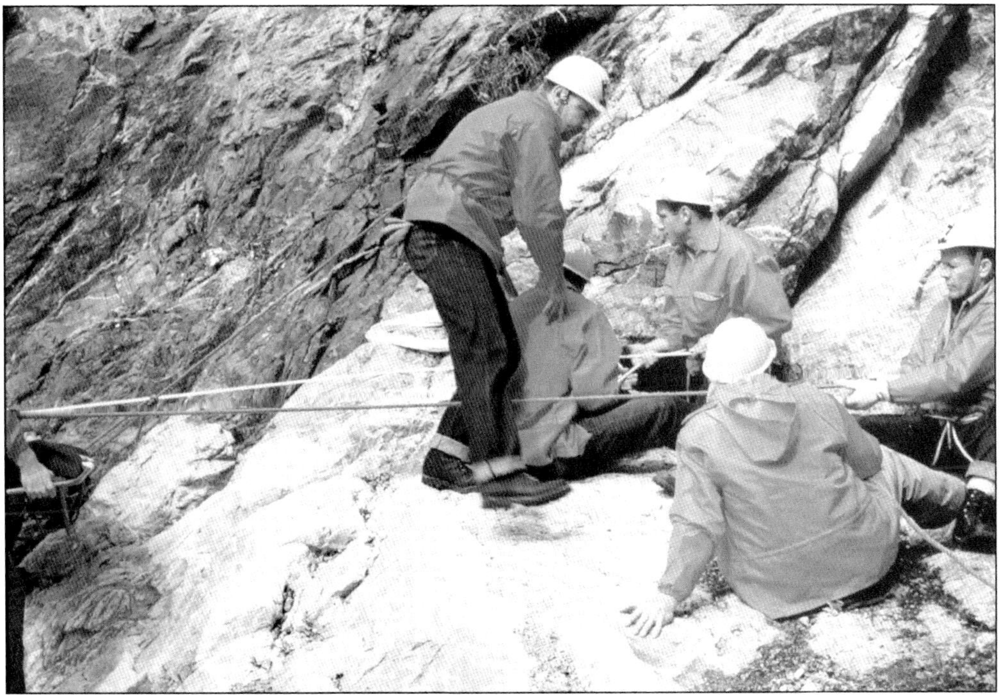

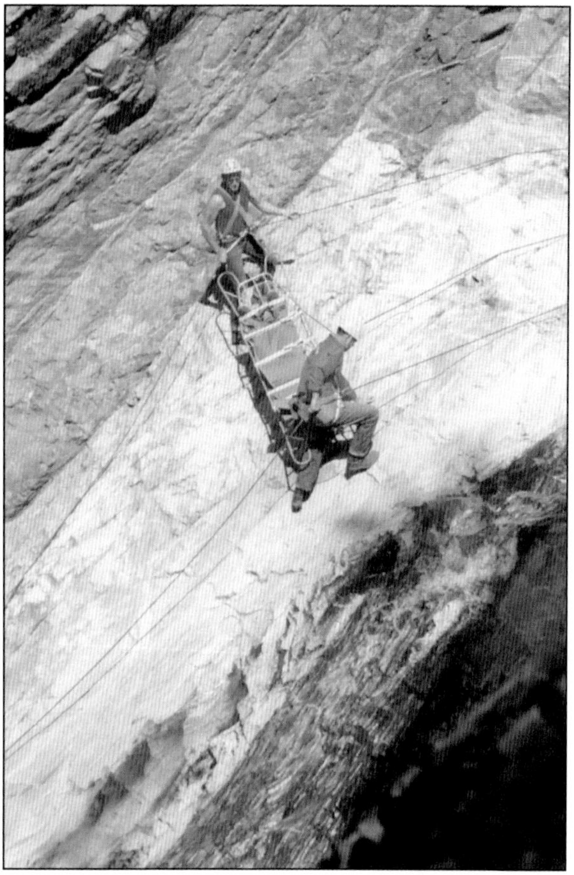

Since the team was founded to fill the need for a qualified rope-rescue team in San Bernardino County, the San Gorgonio Search and Rescue Team has trained extensively in places like Monkey Face Falls in Mill Creek Canyon, near Big Falls. Team members practice lowering a litter down the steep cliffs beside the falls and carrying a litter to another team member at the bottom of the cliff.

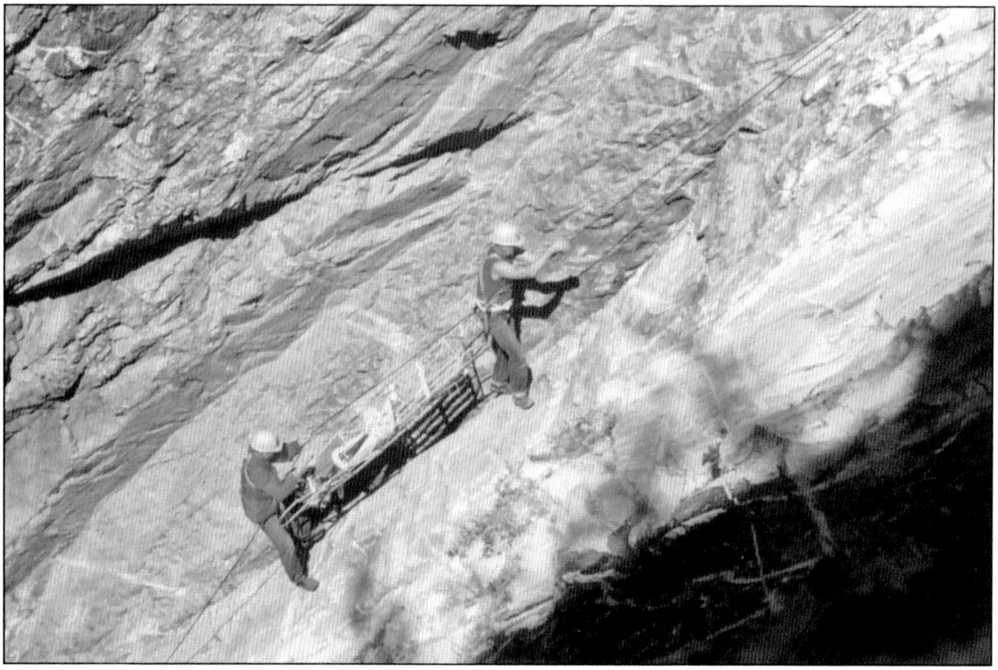

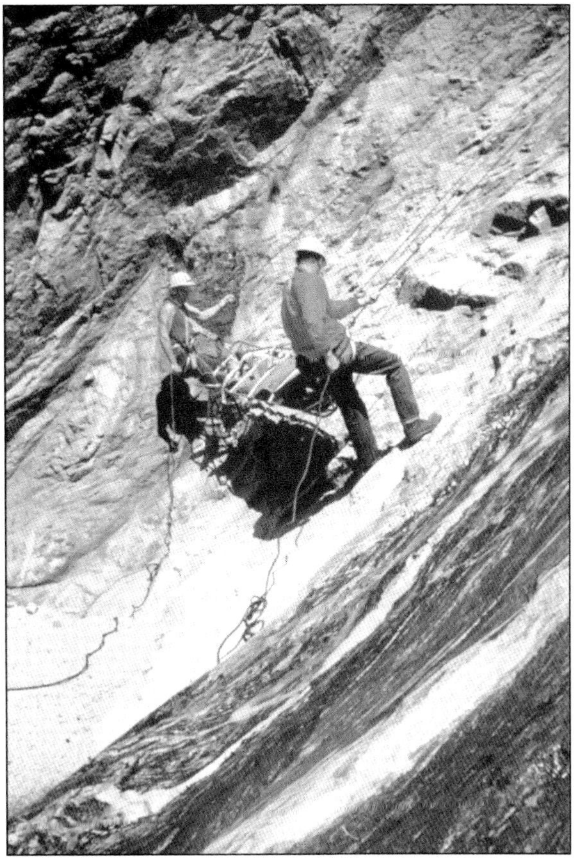

Operating so closely to Monkey Face Falls, team members not only had to be aware of the hazards presented by the steep rock face and loose decomposing granite on one side, but also the dangers of the rushing water and slippery rocks on the other side. Rescuers had only a narrow path to travel to safely bring the litter and the "victim" up and down the rock face.

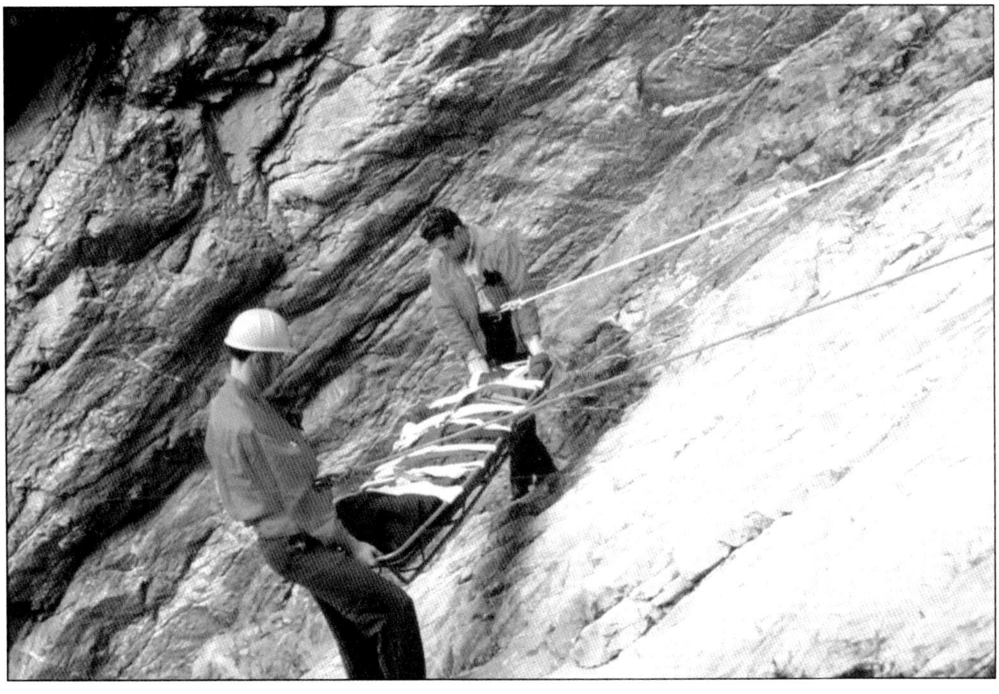

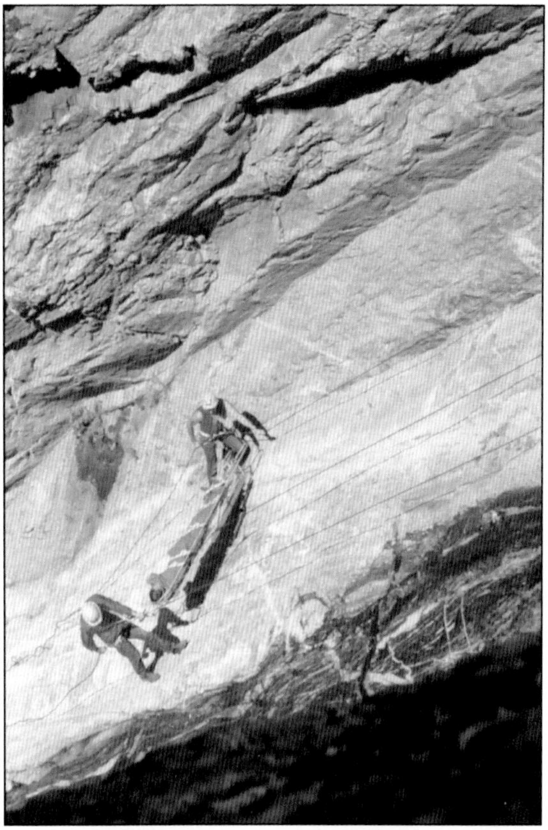

After completing a successful lowering operation with the litter and two rescue team members, the rescuers package the "victim" in the litter and begin the raising operation to take him back to the top of rock face to safety. The rescue team members worked these complicated rope-rescue maneuvers repeatedly. Everyone wanted the chance to play the "victim" and get a ride up the rock face in the litter!

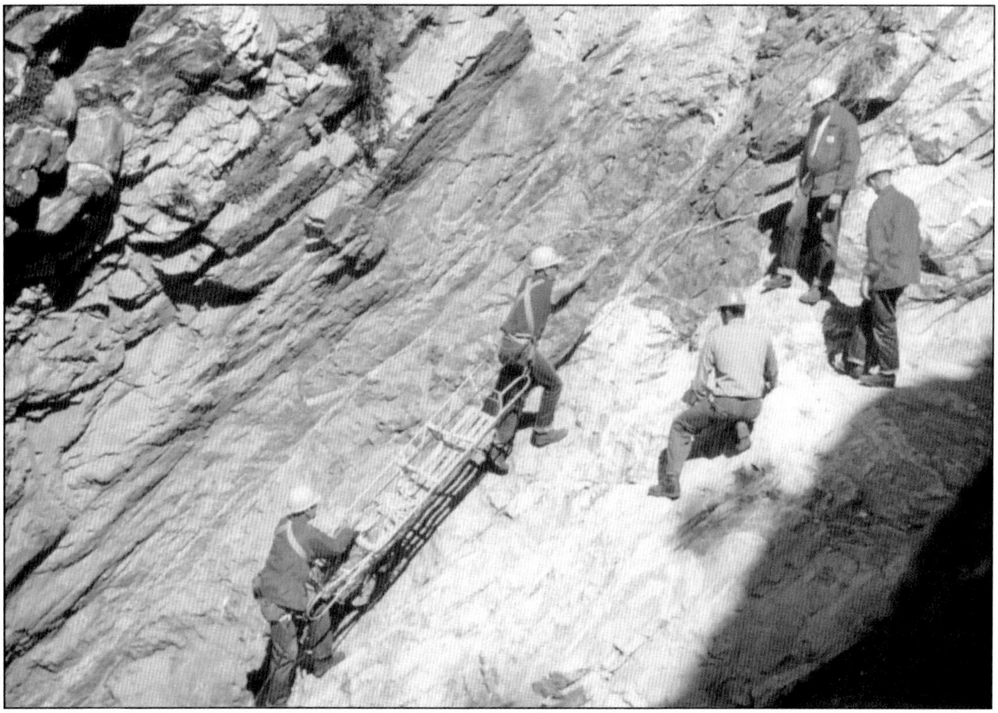

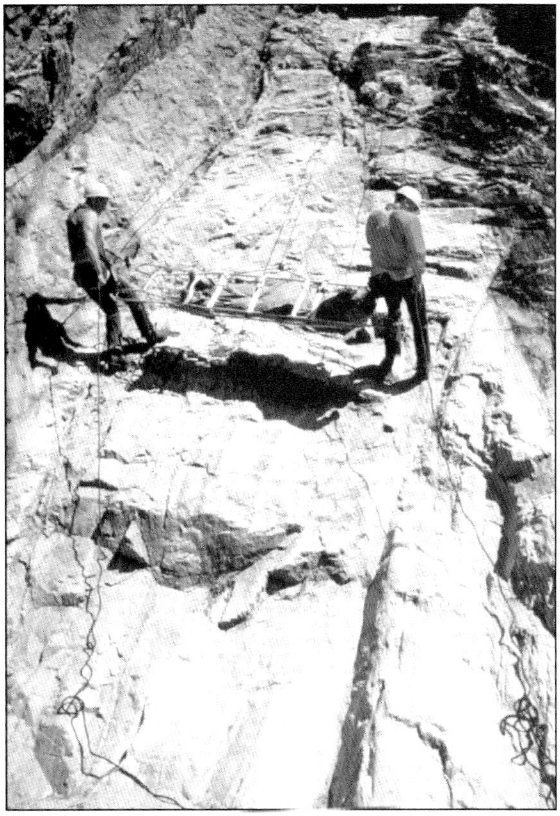

Once the "victim" had been brought up one side of Monkey Face Falls, the rescuers had to lower him and the litter down the other side of the falls to the trail and road below where medical teams were waiting to treat the patient. This lowering took place in several stages because team members had to maneuver around rushing water, large trees, and unstable rock.

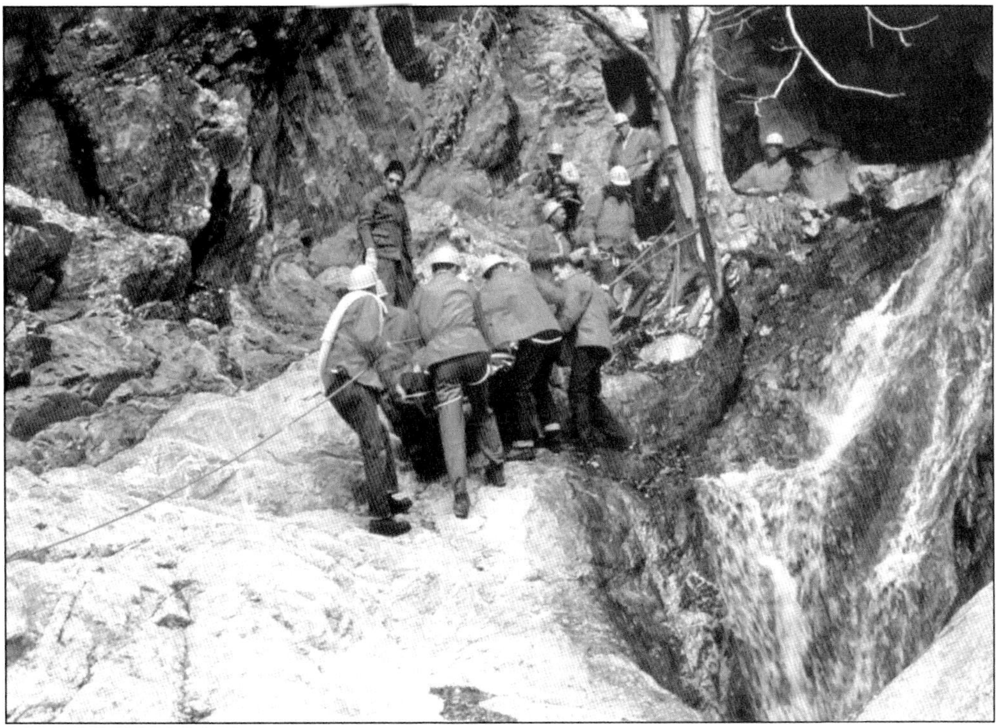

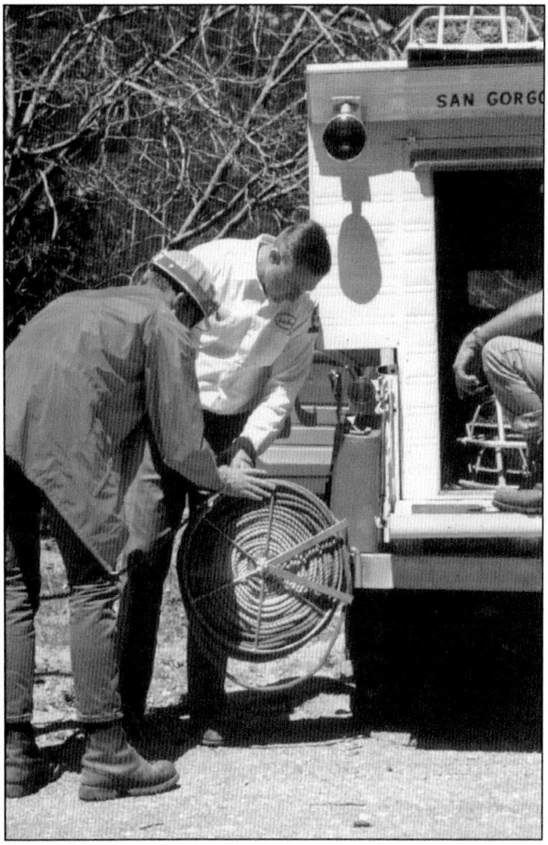

The team quickly realized that they could use their search-and-rescue vehicles as anchors in rope-rescue operations and outfitted them with 500- and 1000-foot reels of rope. The rope could be lowered over the side of mountain roads to aid in the rescue of victims and vehicles that had plummeted off the road.

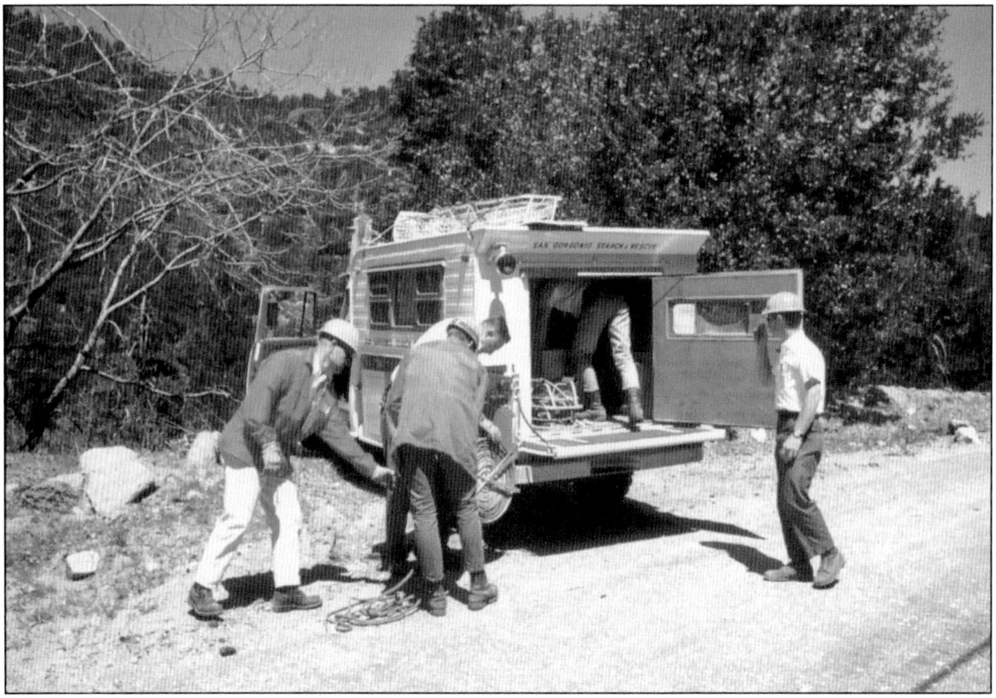

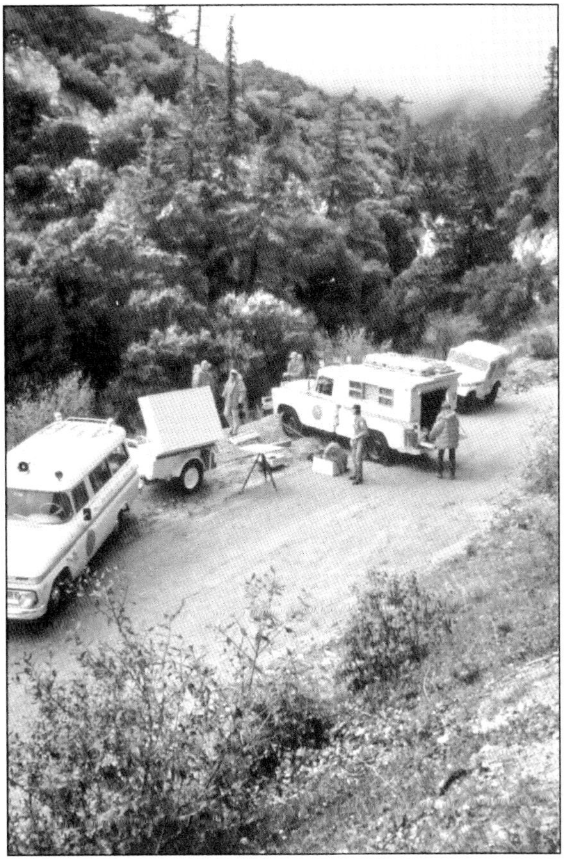

Reinforced with a large steel grill welded to the frame of the vehicle, the front bumper of the vehicle could be used to attach rope-rescue hardware for both lowering rescue team members down to vehicles that drove off the steep mountain roads as well as bring up the litters with the injured and sometimes deceased victims of these accidents.

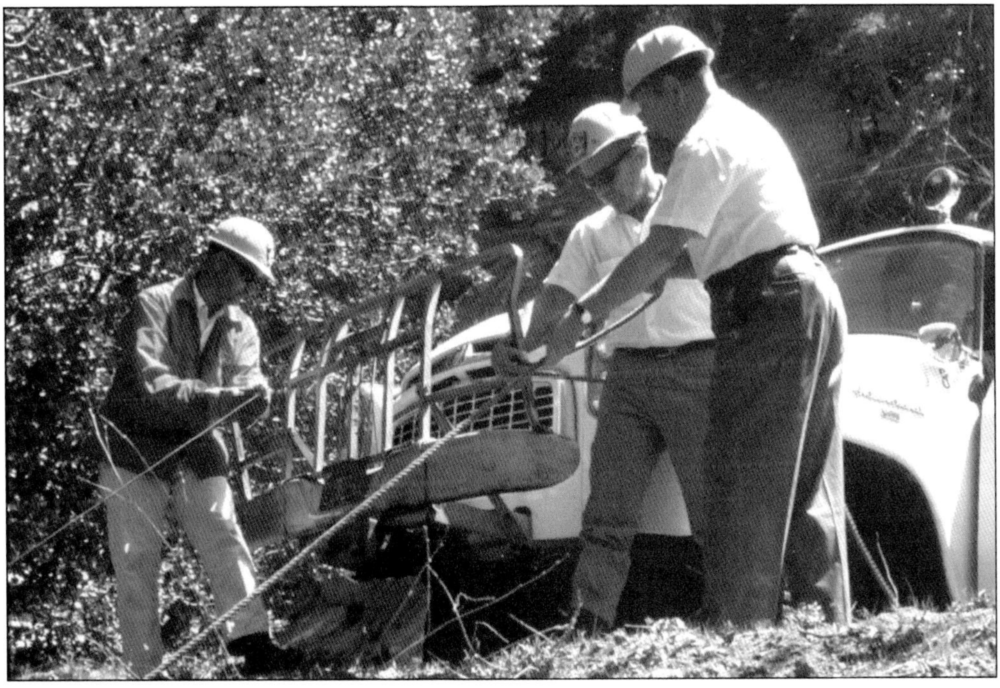

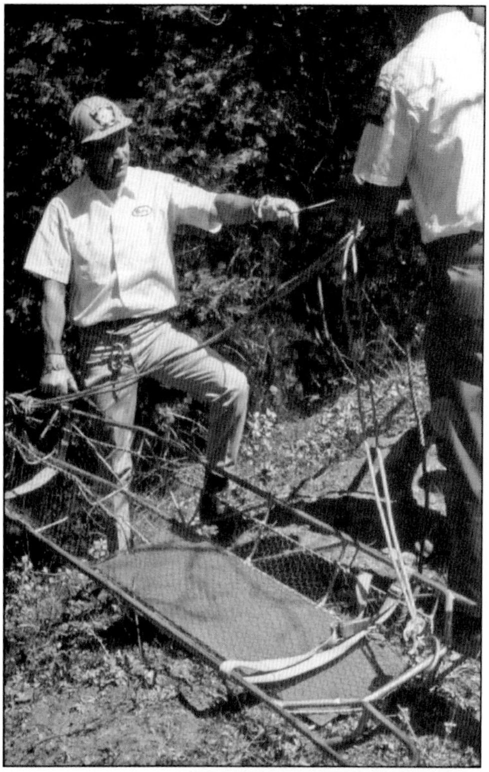

Preparing the litter to go over the side of the road during a training in Forest Falls, team members make sure that the belay and safety lines are attached correctly before attaching the litter attendant to the litter and sending both over the edge.

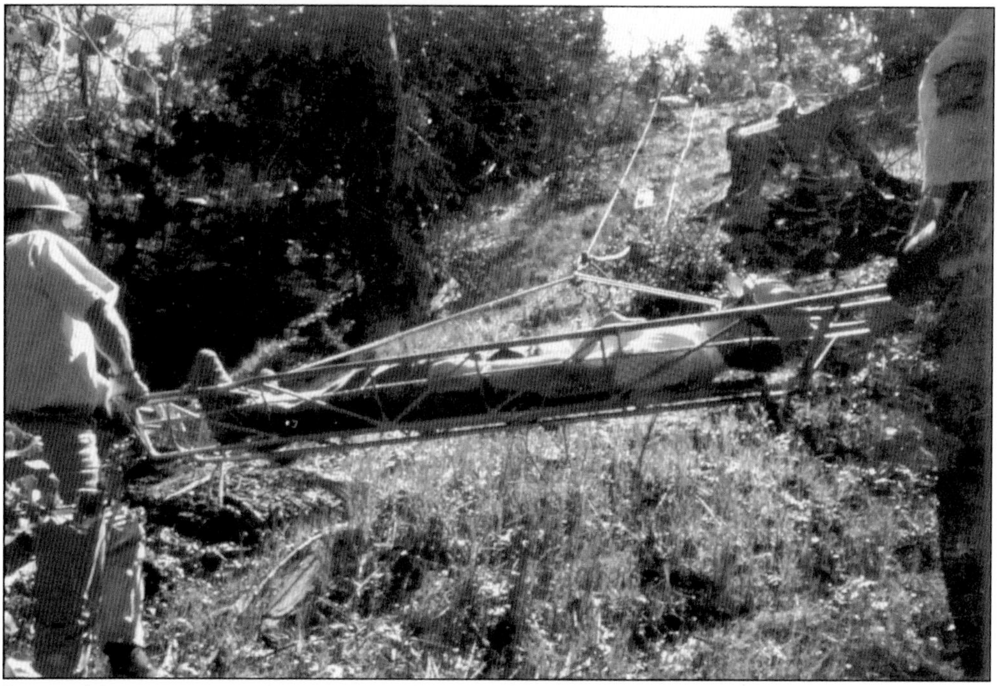

Pictured here is rope-rescue training. Using vehicles as anchors on the side of a road near Angelus Oaks, team members practice raising and lowering a litter with a team member acting as "victim."

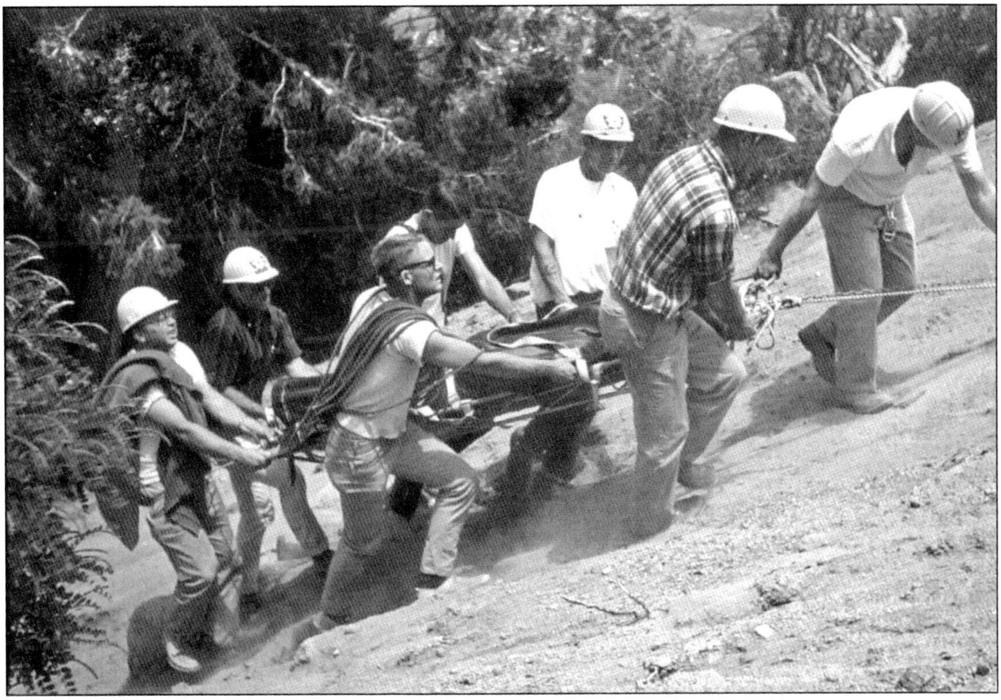

Upon the successful completion of a rope-rescue training session utilizing the team vehicle as an anchor, team members will maneuver the litter over the edge and to the side of the vehicle (out of the "fall zone"), untie the "victim," and start the process all over again with another "victim."

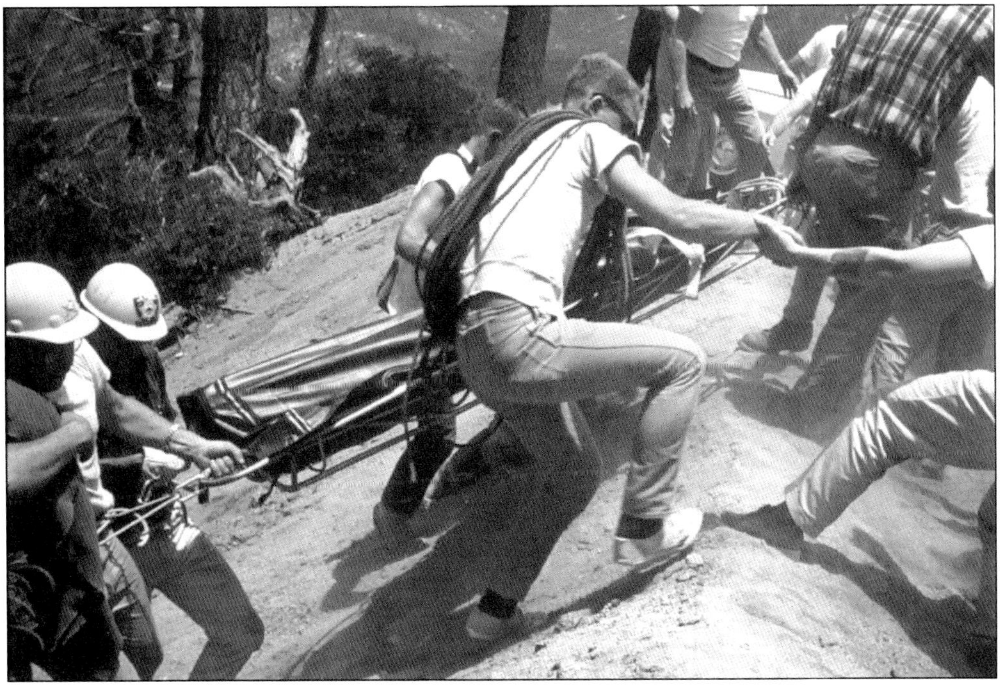

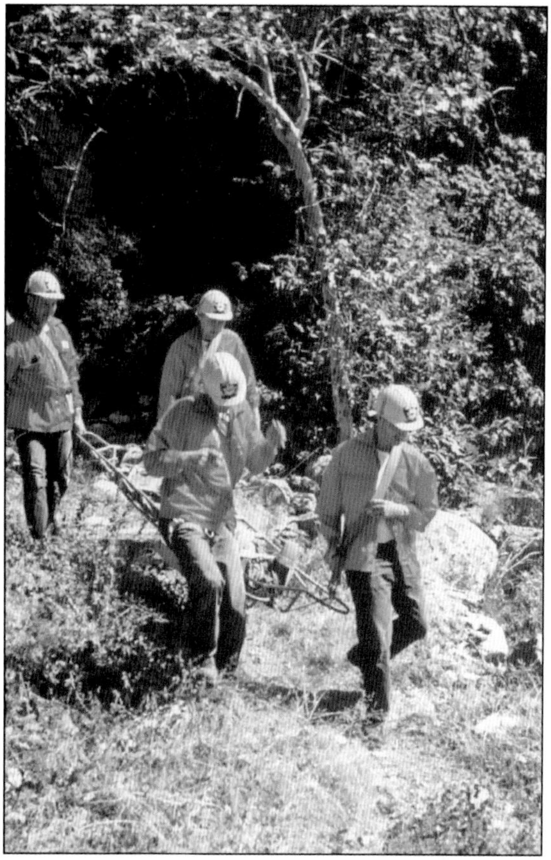

After the successful completion of the rope-rescue operation, team members have to carry the litter by hand to the medical personal waiting at the nearest road. Although the rope-rescue operation over the side of a road or a cliff is the more technical part of the rescue, the carry out of the litter is by far the most physically challenging part of any recovery.

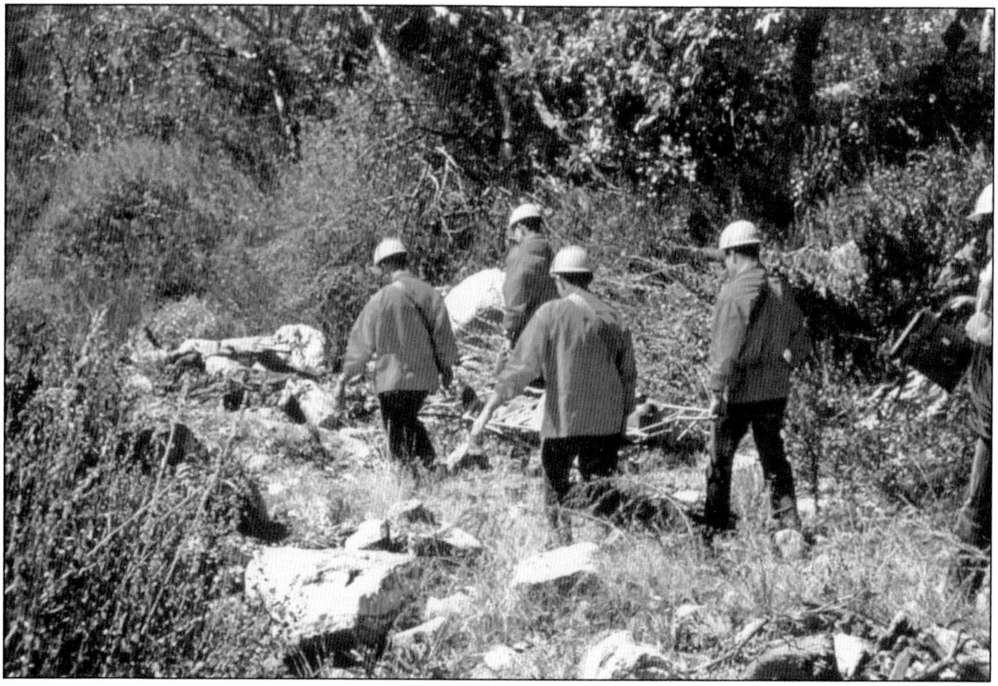

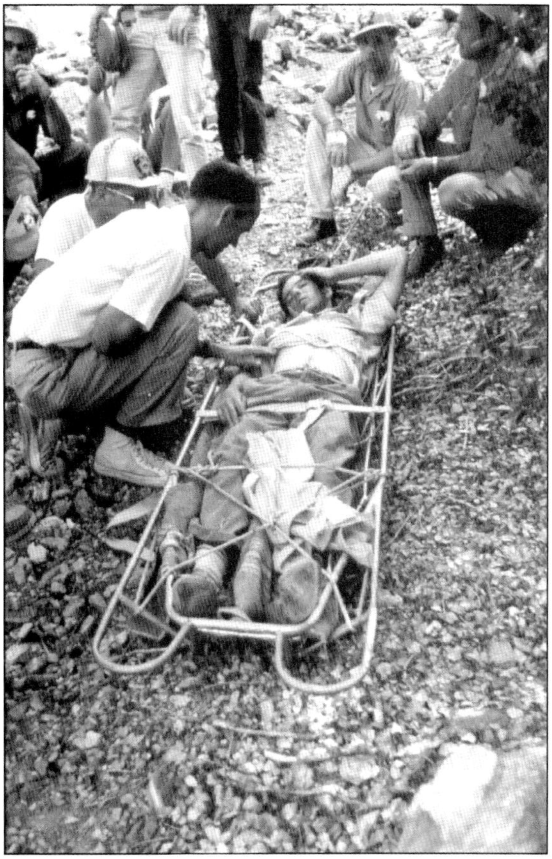

Many times there is not a clear trail to follow and rescuers have to maneuver around rocks and brush to get the victim to safety. Over the years, it seemed that the more remote the place, the more severe the injuries of the victim were. No one ever gets seriously injured next to a road or trail where the rescue would be easy.

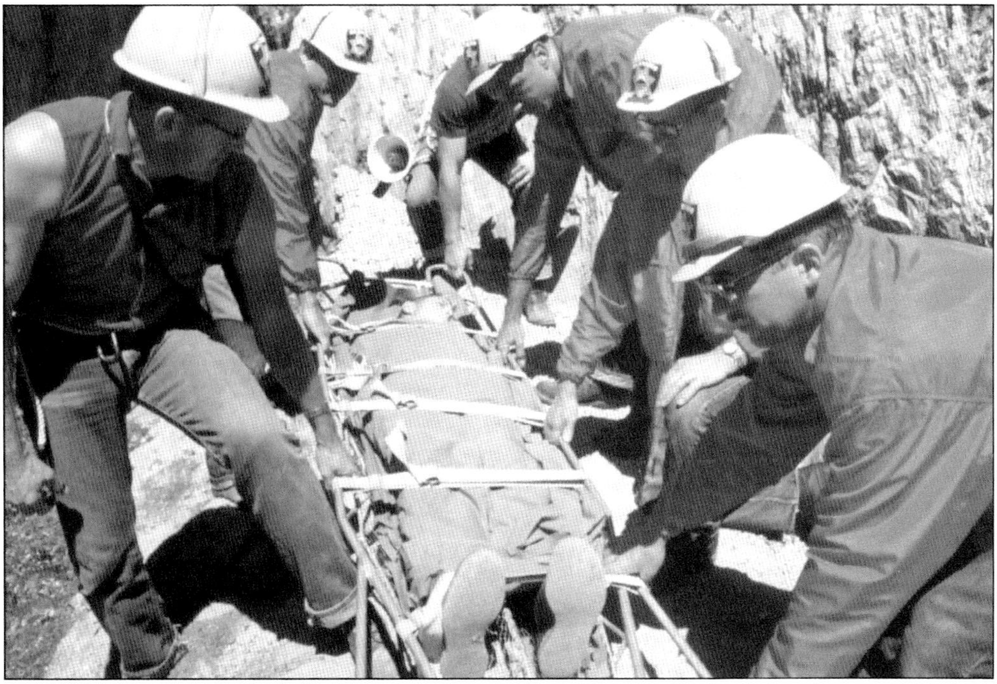

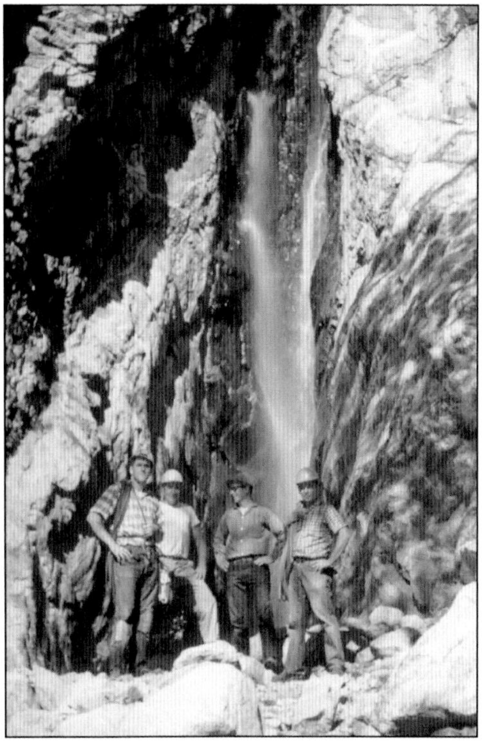

Team members Ed Walker and Fred Couch pose are included in this picture, which was taken at the base of Monkey Face Falls after a long day of team rope rescue training The newly formed rescue team did not have a standard uniform shirt or helmets.

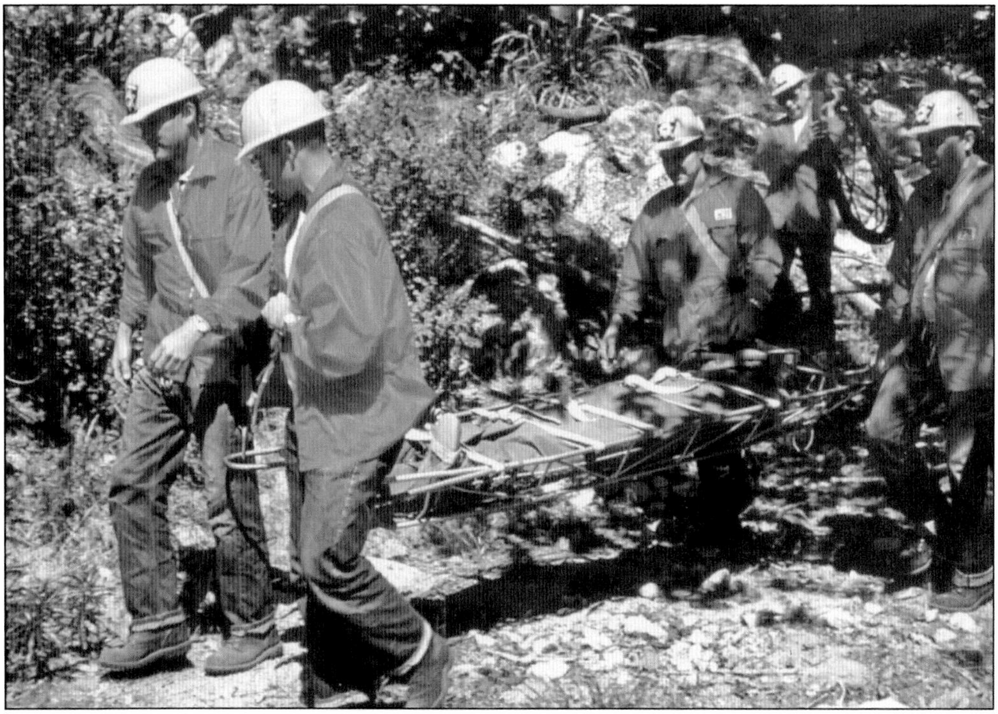

Team members used shoulder straps to take some of the weight from the litter off their arms and distribute it to the rest of their body. They would also alternate carrying the litter with carrying other team equipment, allowing team members to have a rest from carrying the heavy litter.

Here team members inspect the base of Monkey Face Falls for a fallen climber. The climbers rope can faintly be seen though the falls. The climber, who had not fallen all the way to bottom of the falls, was later discovered hanging approximately 60 feet above the rescuers.

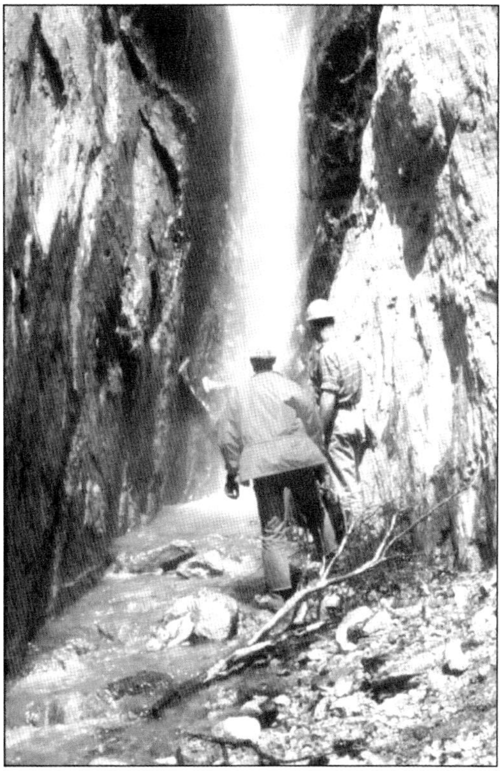

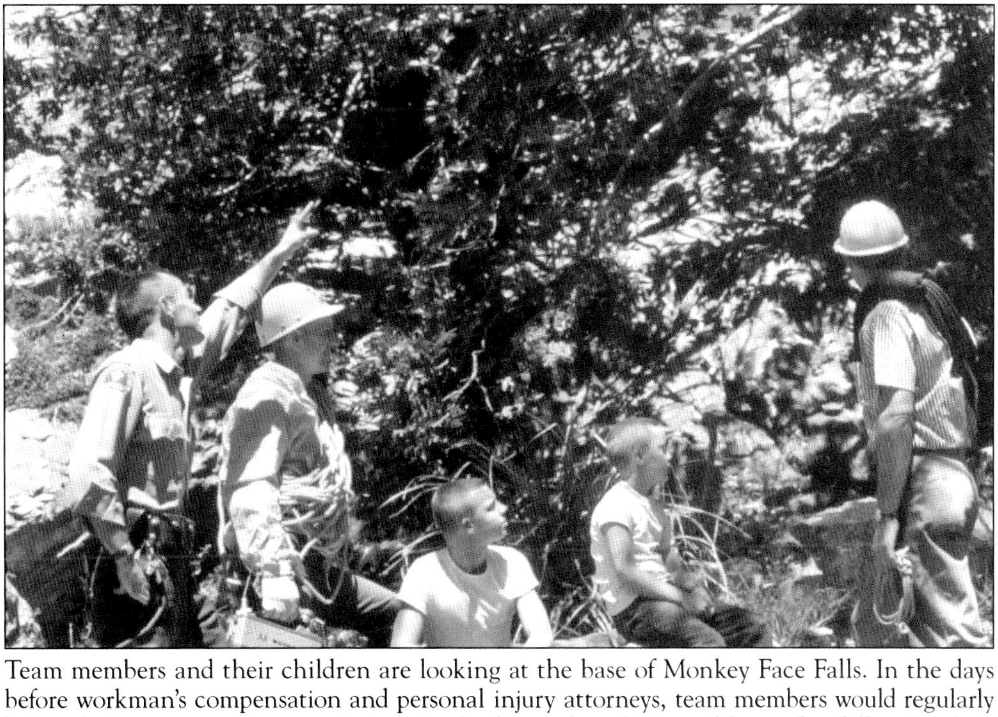

Team members and their children are looking at the base of Monkey Face Falls. In the days before workman's compensation and personal injury attorneys, team members would regularly bring their teenage sons to trainings in the hope that they would one day follow in their footsteps as rescuers.

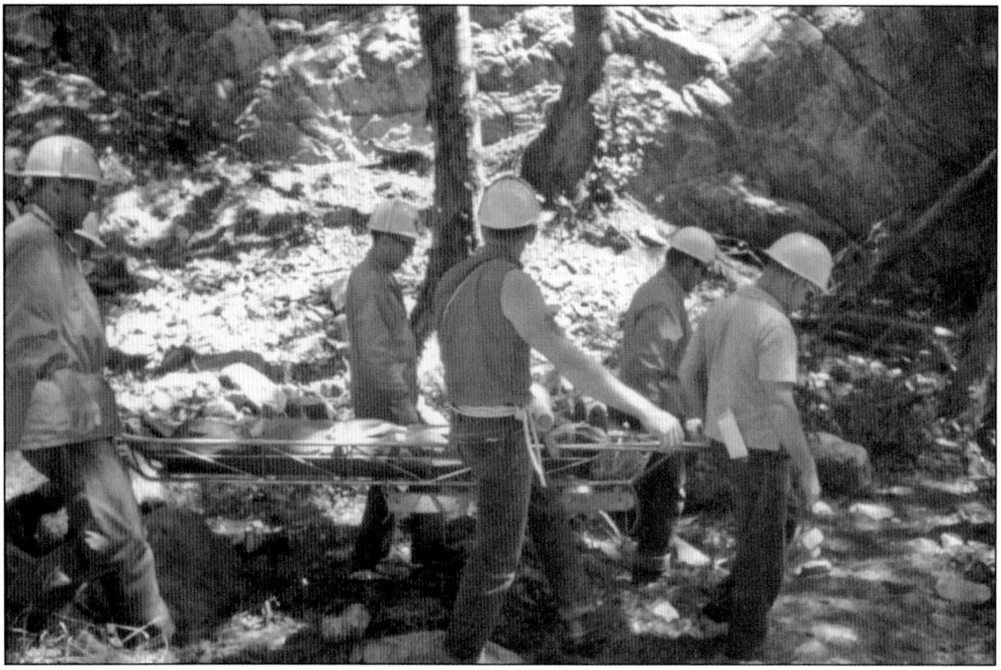

Even though the rescuers are off the steep rock face and the ropes are put away, they still have the arduous task of carrying out the injured climber and getting to medical attention as soon as possible. Though signs have been posted for decades explaining the danger of climbing Big Falls in Mill Creek, many people still climb every year and a few get hurt or killed on the dangerous waterfall.

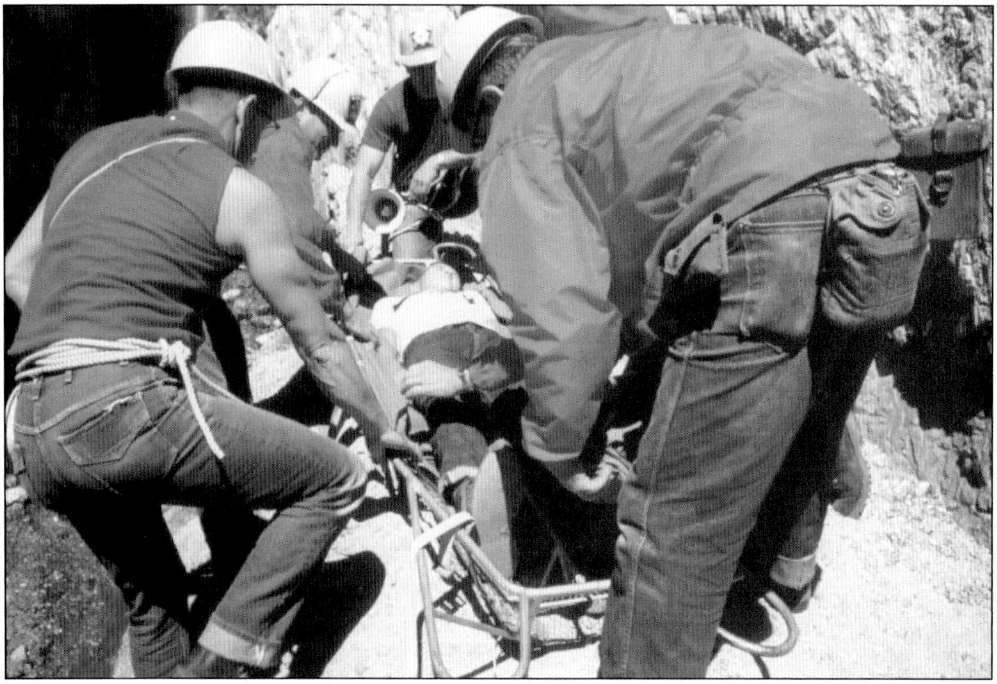

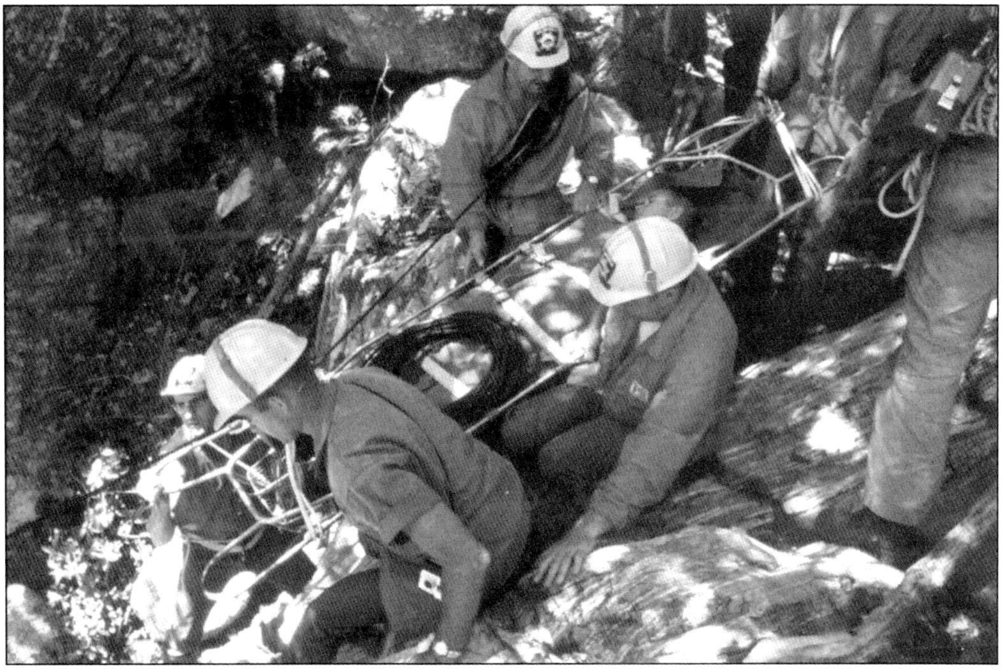

Upon the completion of successful rope rescue of an injured climber on Big Falls in Mill Creek in 1962, team members have to negotiate the litter, carrying the injured around boulders and thick brush to get him to the medical personnel waiting at the nearby road.

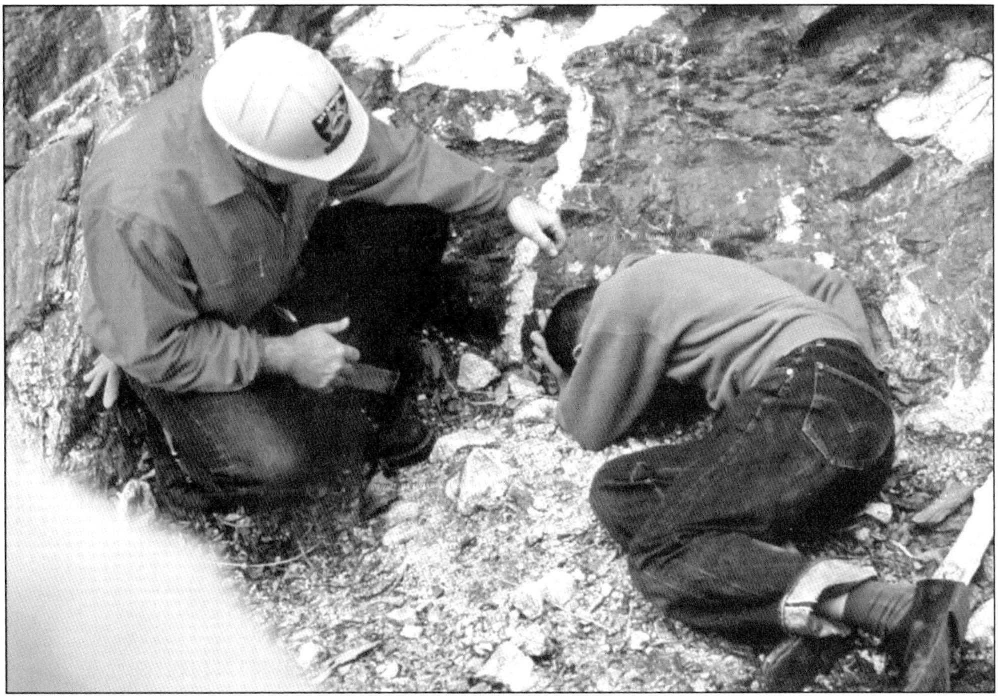

Thom Mayfield makes contact with a fallen climber at the base of Big Falls in Mill Creek Canyon. All team members have some degree of medical and first aid training. The team now has a standardized jacket and helmet as part of their uniform.

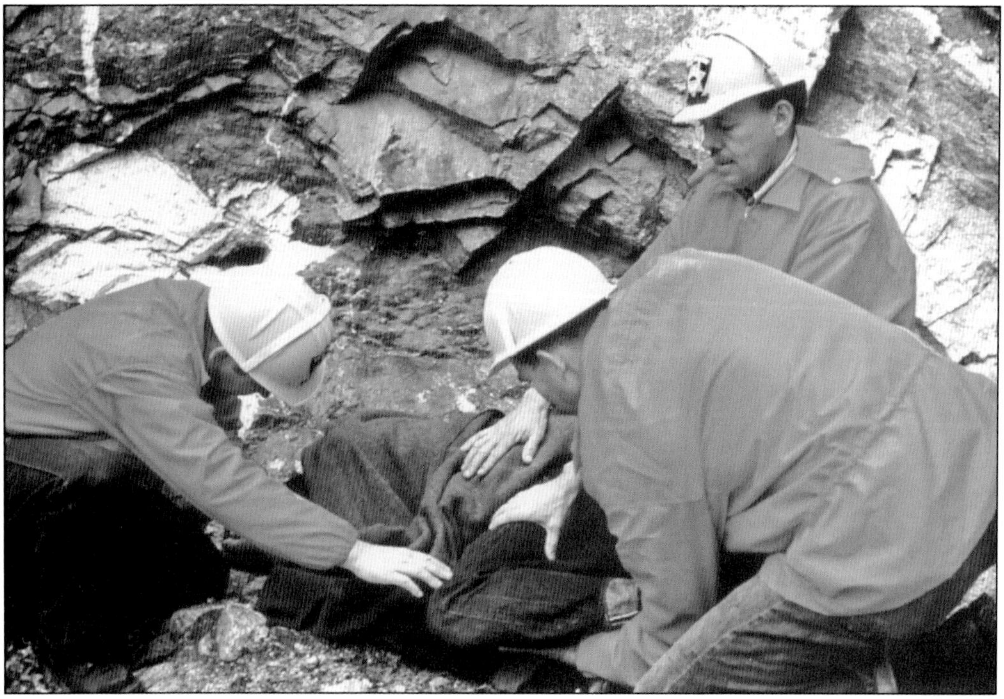

Thom Mayfield and Barry Lee assess the injuries of a fallen climber at the base of Big Falls in Mill Creek Canyon. Rescuers put a blanket on him to keep him from going into shock and try to move him into a position where they can put him in a litter and carry him out to waiting medical personnel.

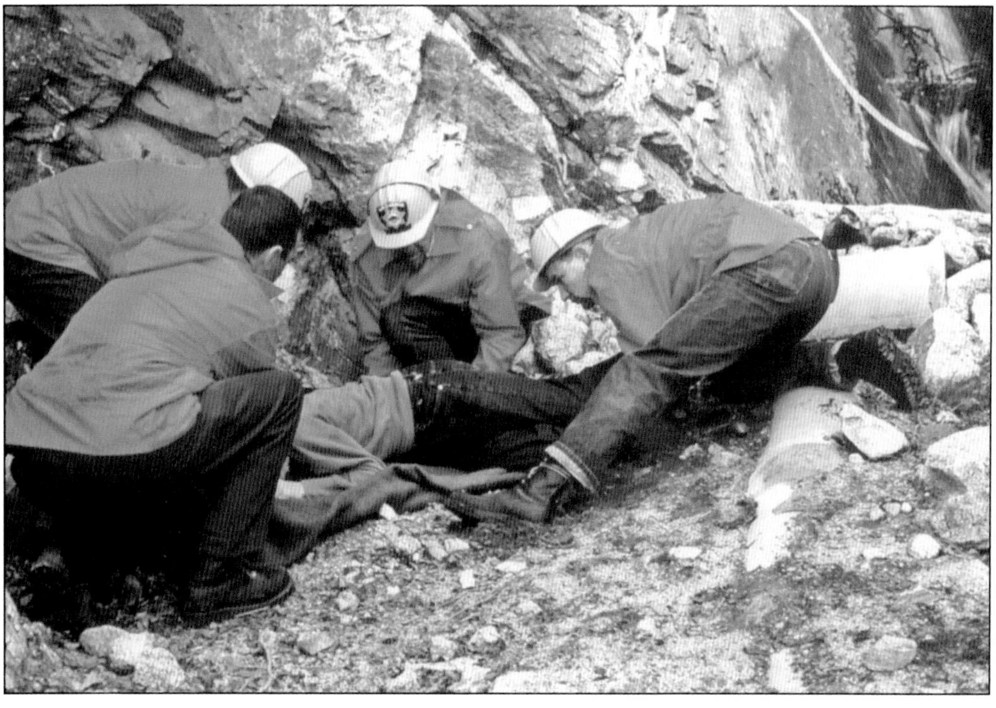

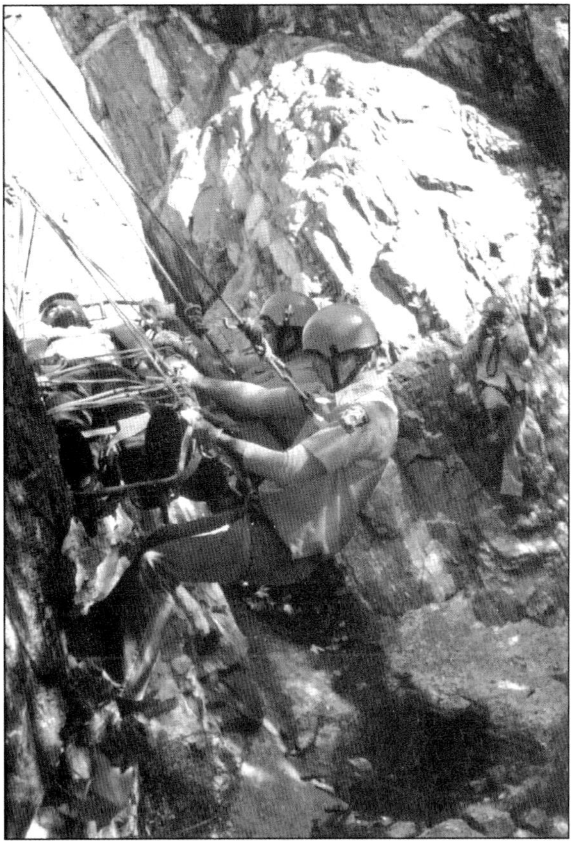

Though the team uniforms and helmets have changed, many of the rope-rescue procedures have stayed the same. Team members are still training on near vertical rock faces and practicing the safe lowering and raising of "victims" in the litter, waiting for the call some weekend when they will have to rescue an actual fallen climber on these dangerous rock faces.

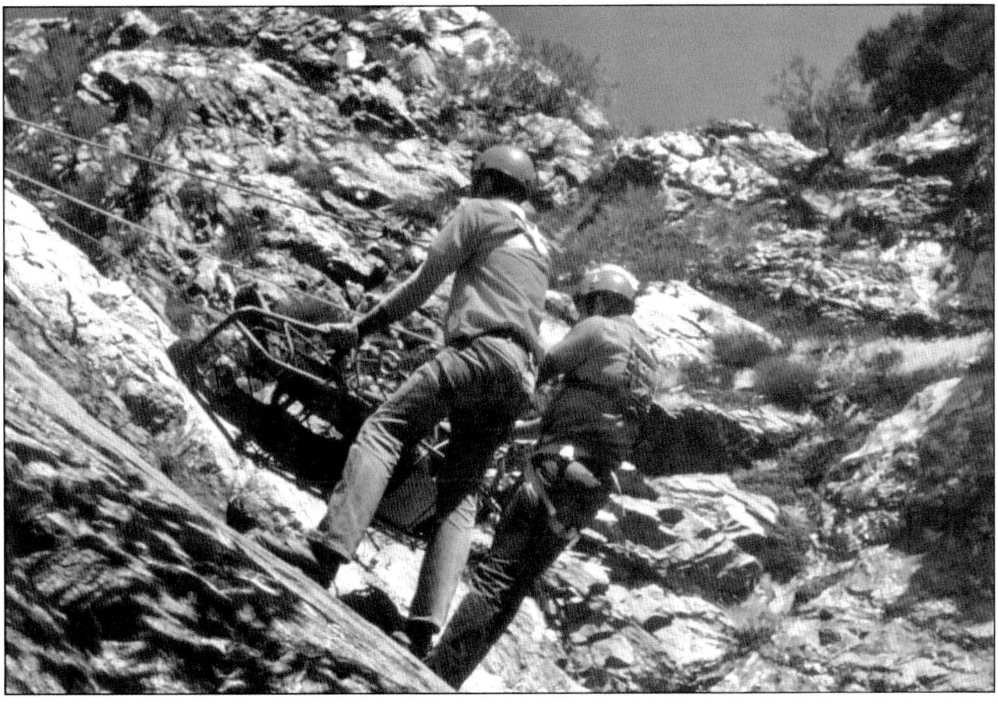

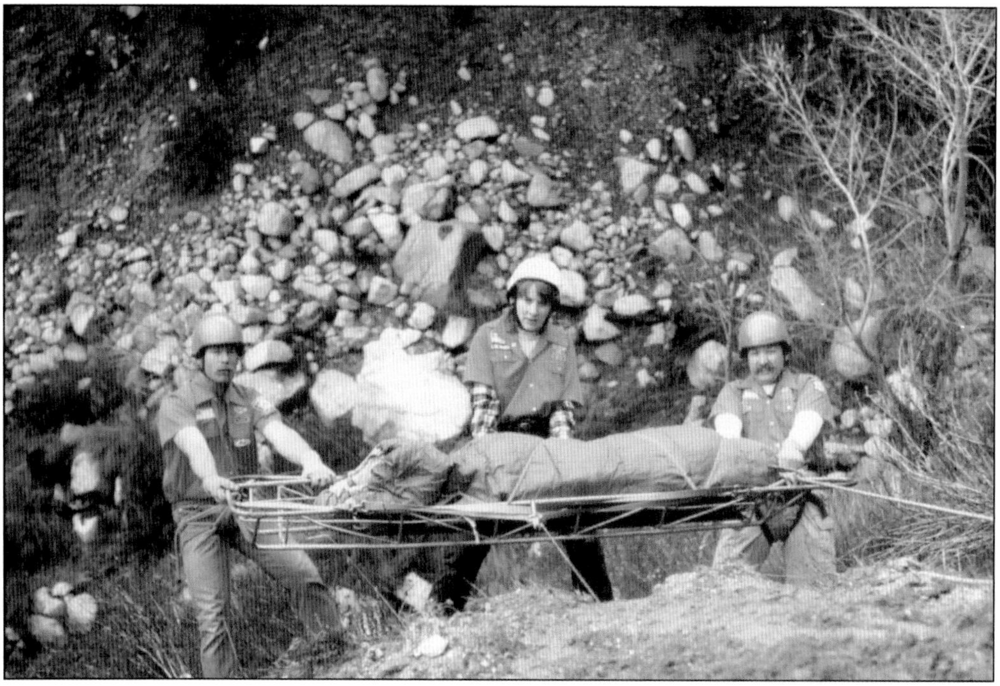

Team members Dean Caldwell (left), Ellyn Loenhorst (center), and Skip Brown bring up the litter with the victim of a fatal car accident off State Highway 38 in Angelus Oaks. Ellyn was the first woman to join the team, in 1985. This caused a little bit of controversy among some of the wives of the other team members. Since Loenhorst, 11 more women have joined the team, and all of them have proven that when it comes to search and rescue, gender is not an issue.

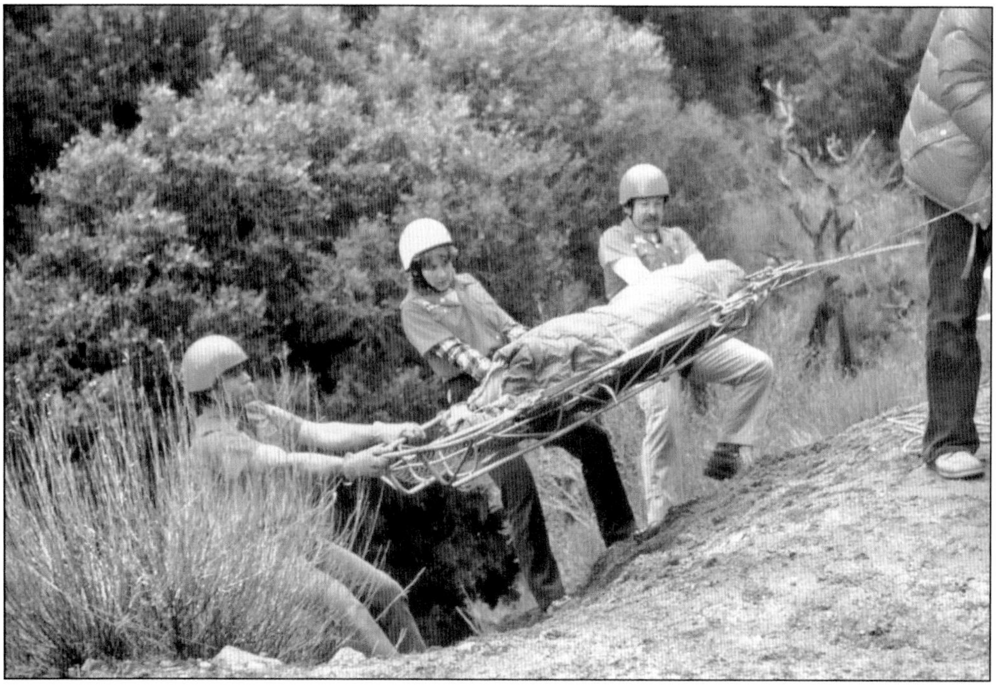

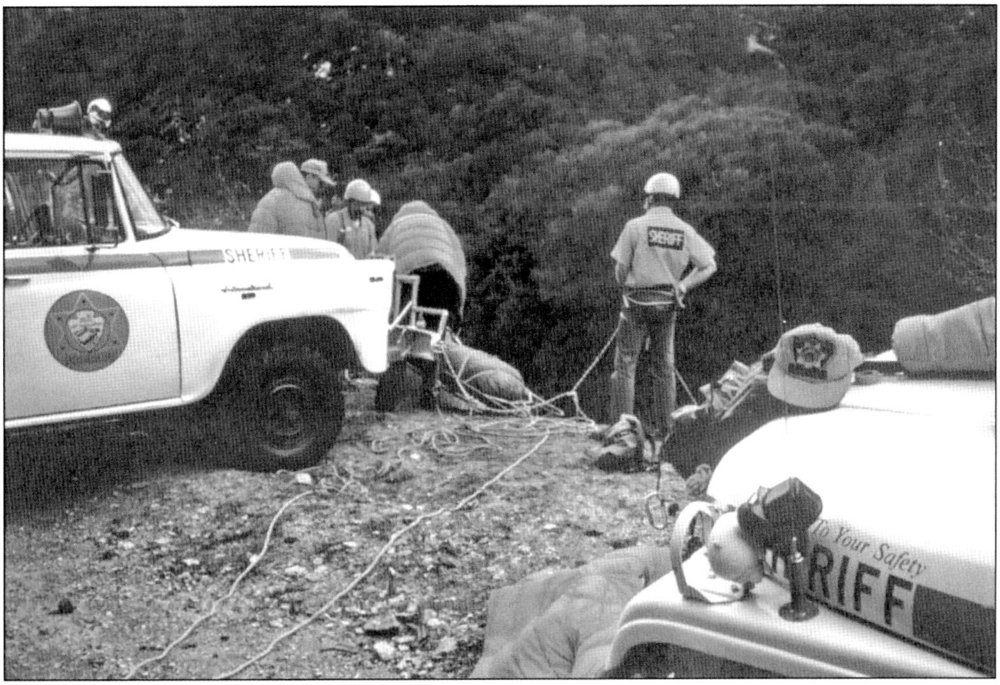

Here team members train by using the front end of a search-and-rescue vehicle as an anchor. Though the vehicles and uniforms have changed over the years, rope-rescue procedures did not change for several decades.

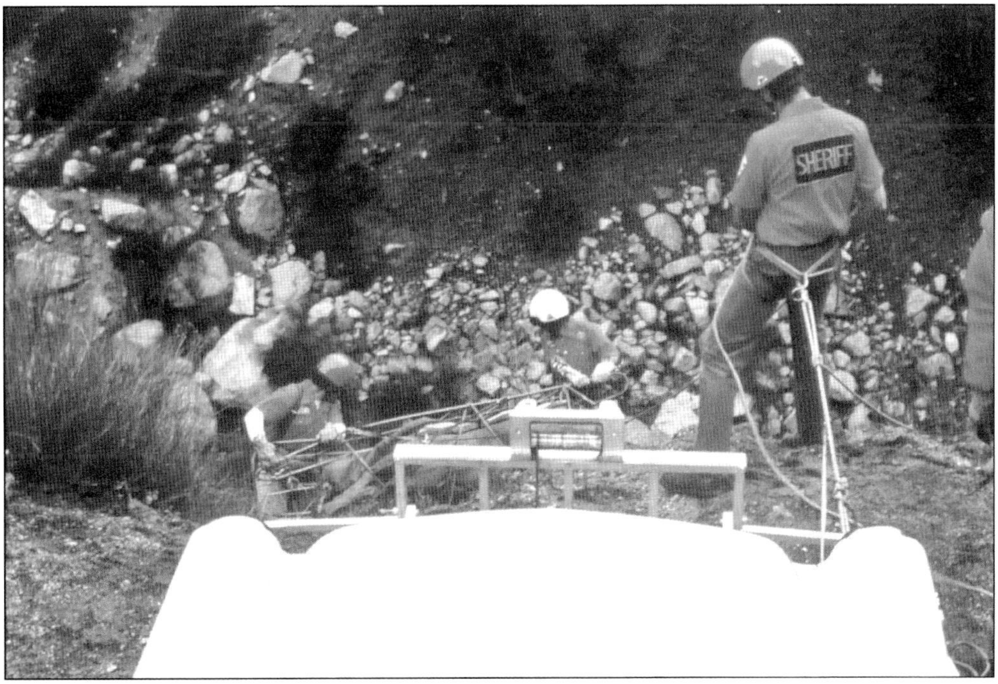

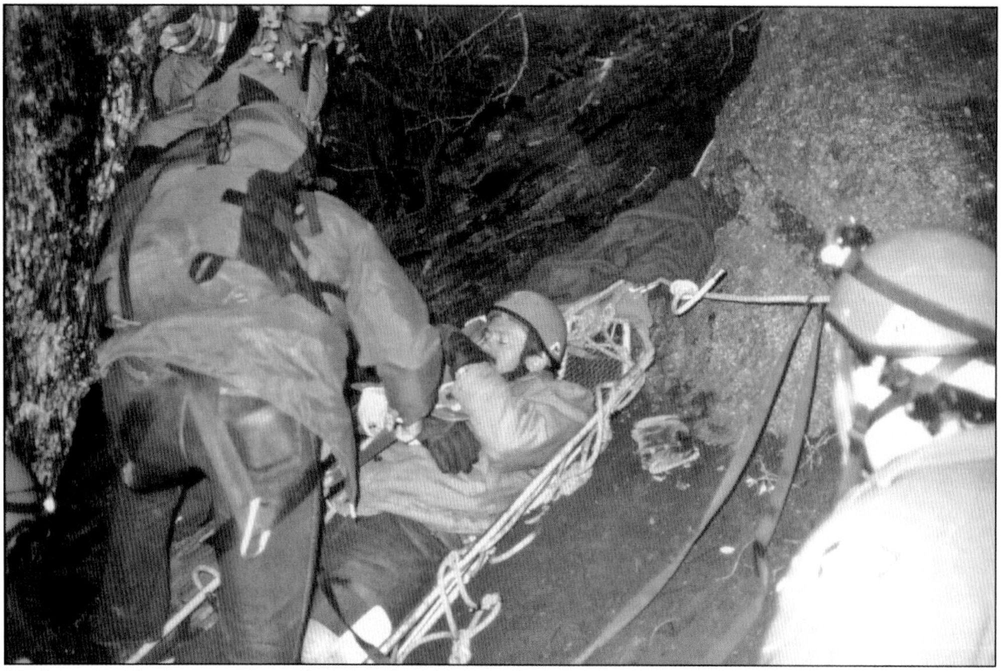

Unfortunately, most rescues do not take place in the middle of the day. Here, late into the night, rescuers work to splint the broken leg of an injured hiker, strap him into the litter, and begin the long process of hauling him back up the steep hillside where he fell on the Vivian Creek Trail.

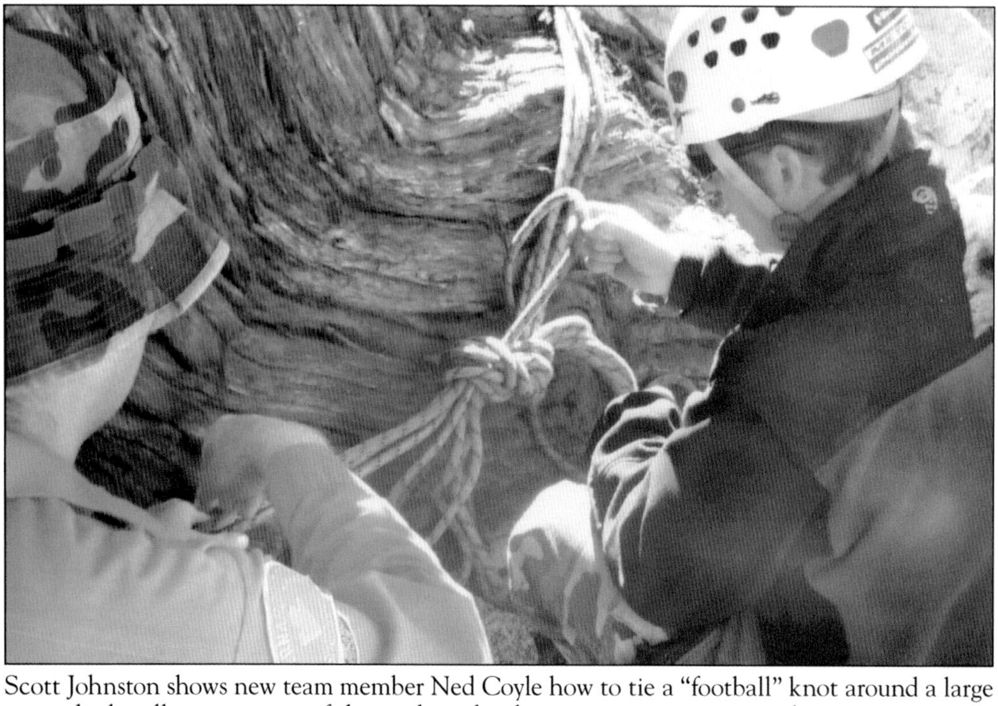

Scott Johnston shows new team member Ned Coyle how to tie a "football" knot around a large tree, which will serve as one of the anchors for the rope-rescue operation the team is training for. The "football" knot is a variation of the figure eight knot and gets its name because when tied properly it looks like football.

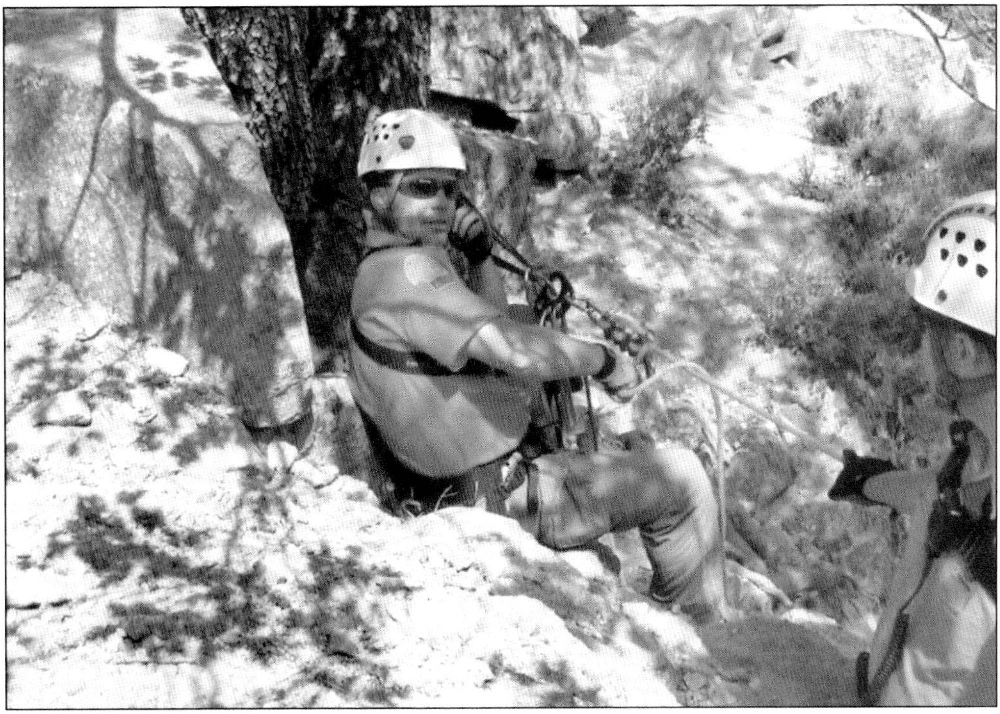

Larry Reinhart and Shane Elliot are manning the main-line system. The main line is used to raise or lower the load (a victim and the litter) up or down a cliff to get them to safety. The main line is lowered using a steel brake bar. The main-line operator can easily adjust the speed of the main line with this bar while keeping it secure.

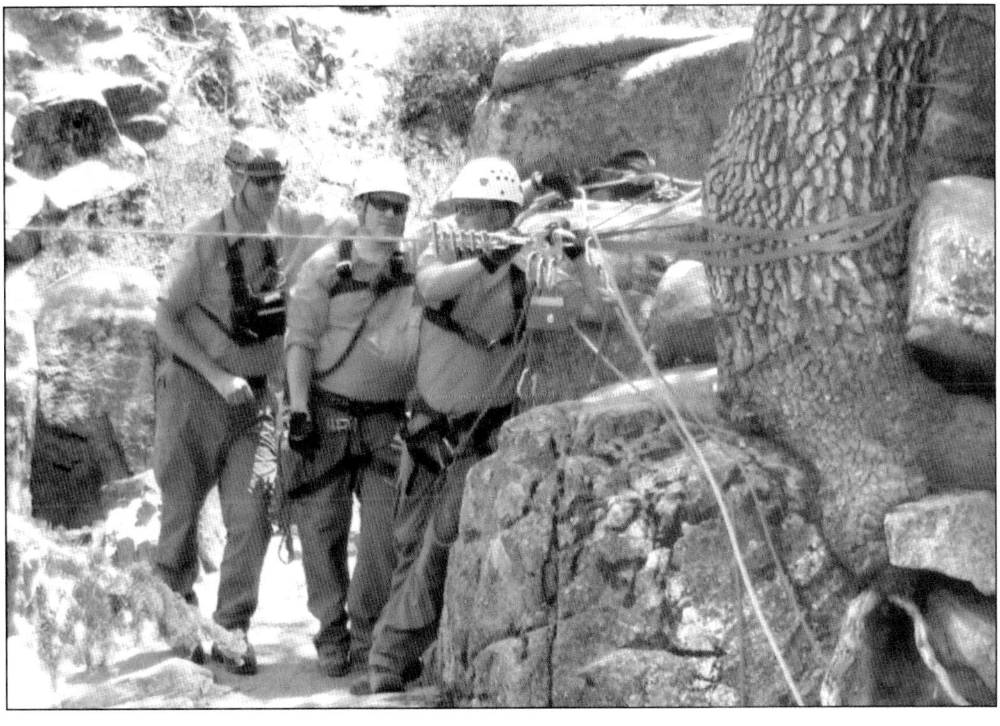

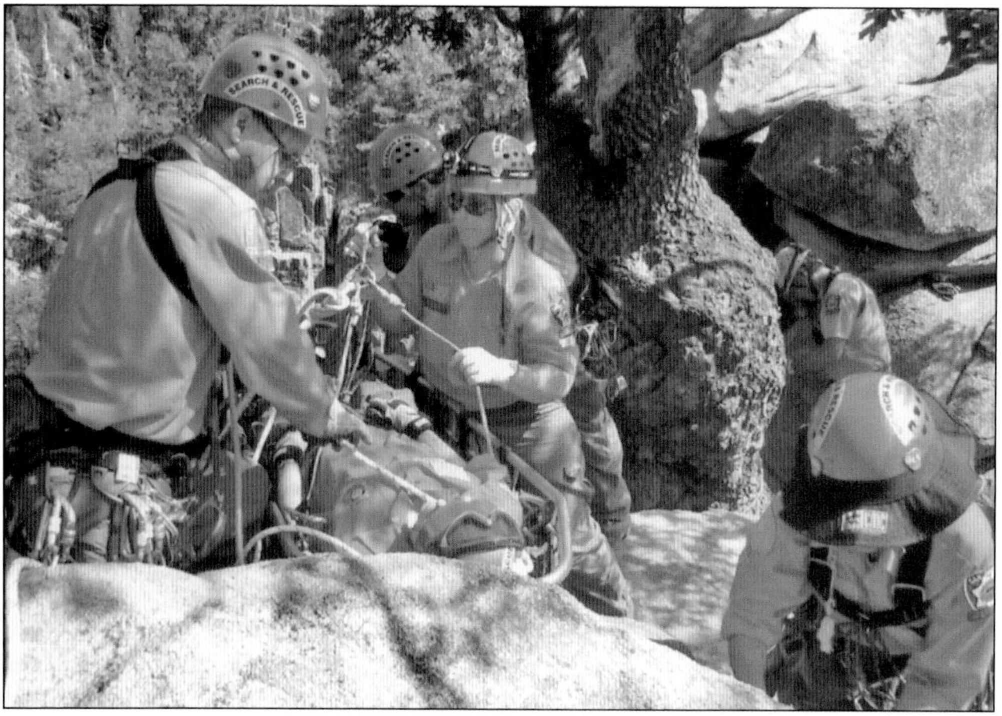

The litter is prepared to go over the side of steep cliff in Deep Creek near Running Springs, California. Deep Creek is one of the team's favorite places to train for rope-rescue operations.

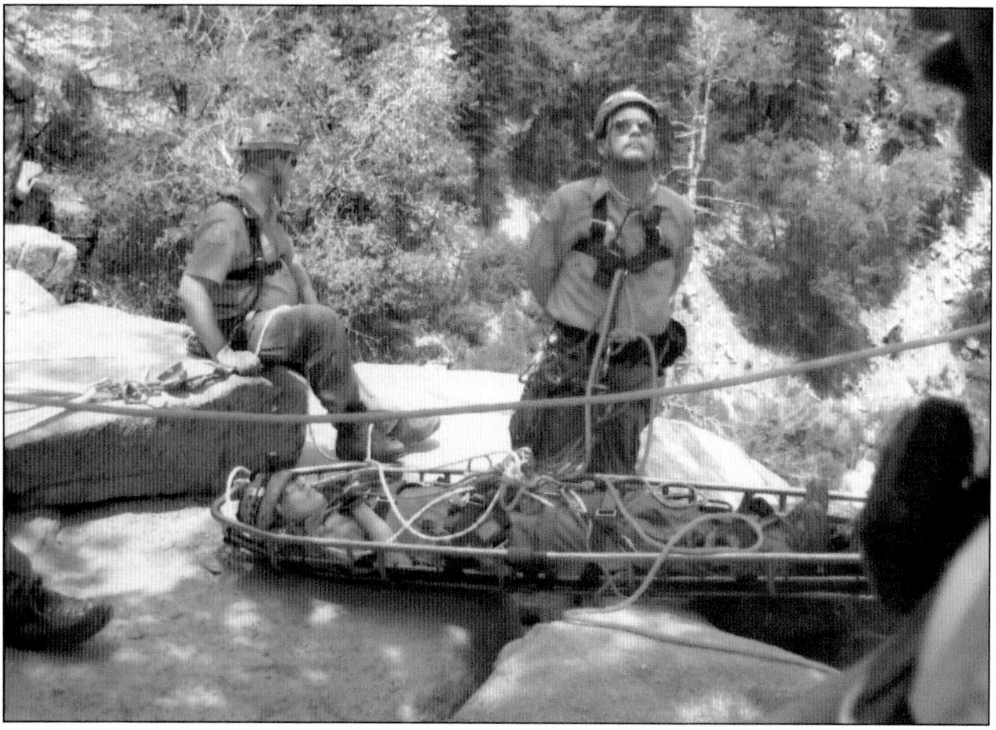

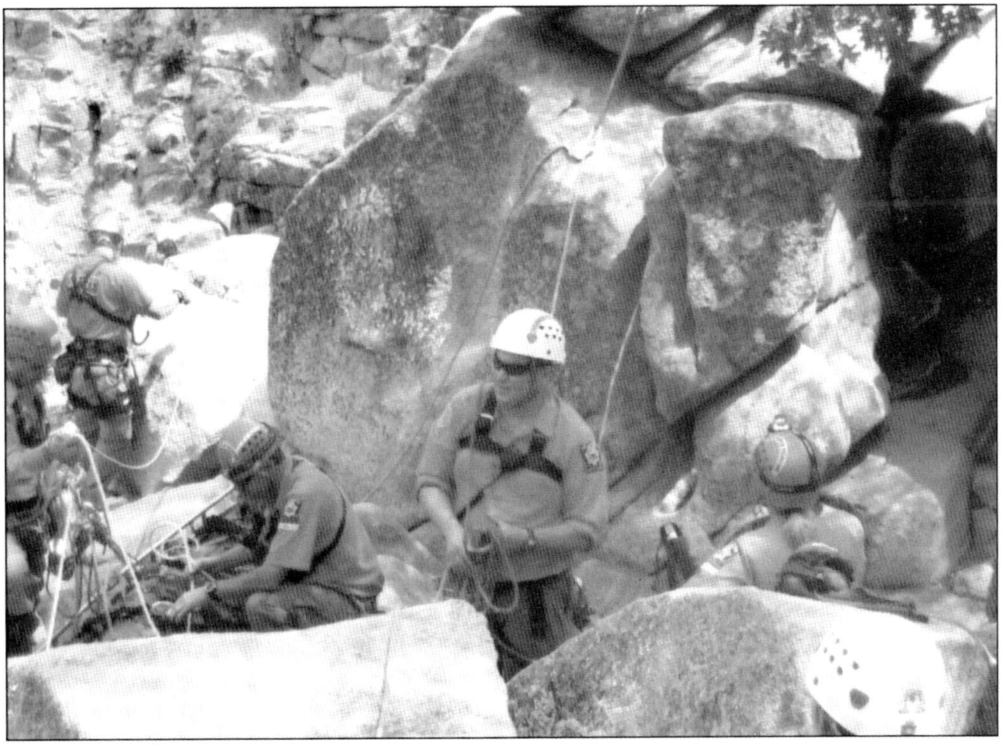

After successful rope-rescue training, team members disassemble the litter rigging and prepare for the next training exercise. The team uses its own members as mock victims; it is typically a rite of passage for the newest members of the team to play that role.

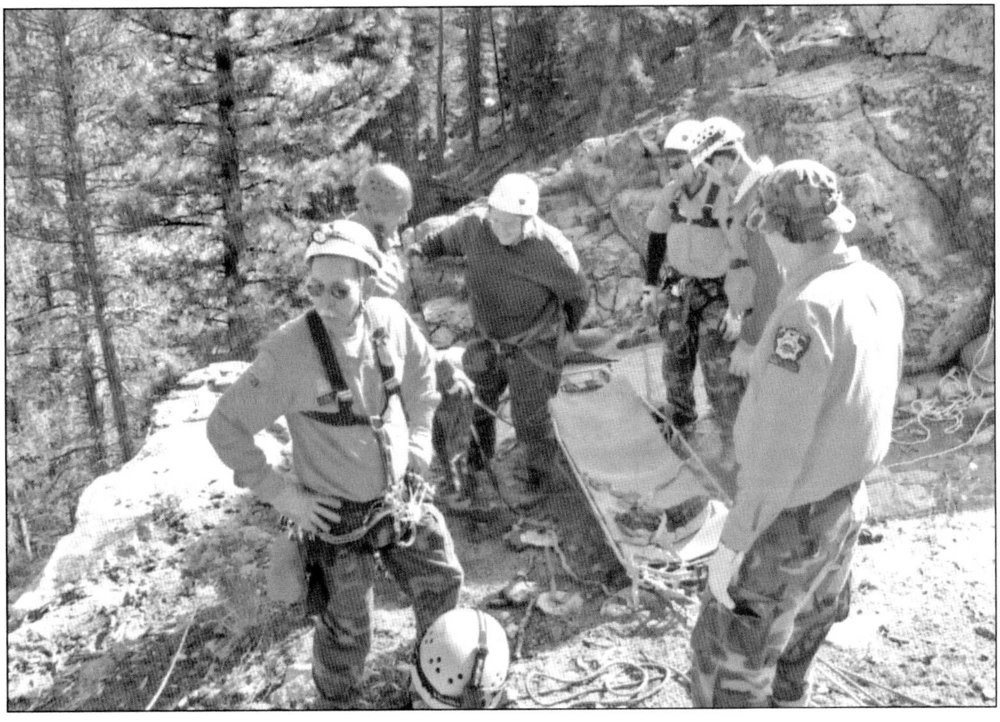

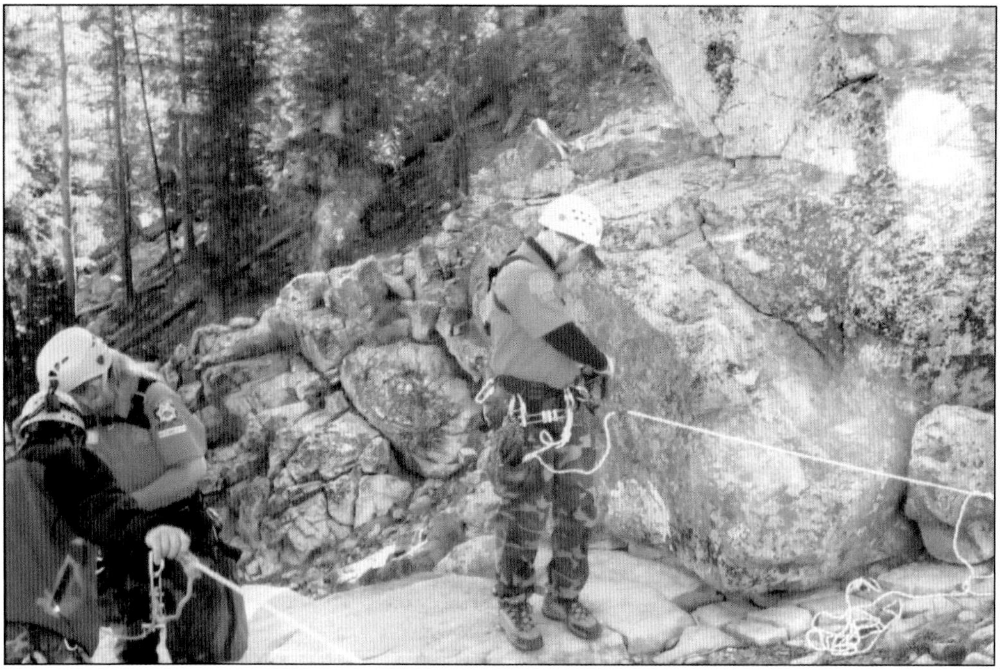

Team members get ready to rappel down the side of rock face near Onyx Summit off Highway 38 near Big Bear, California. Since the early days of the team, rappelling has always been a basic skill that every team member is expected to have.

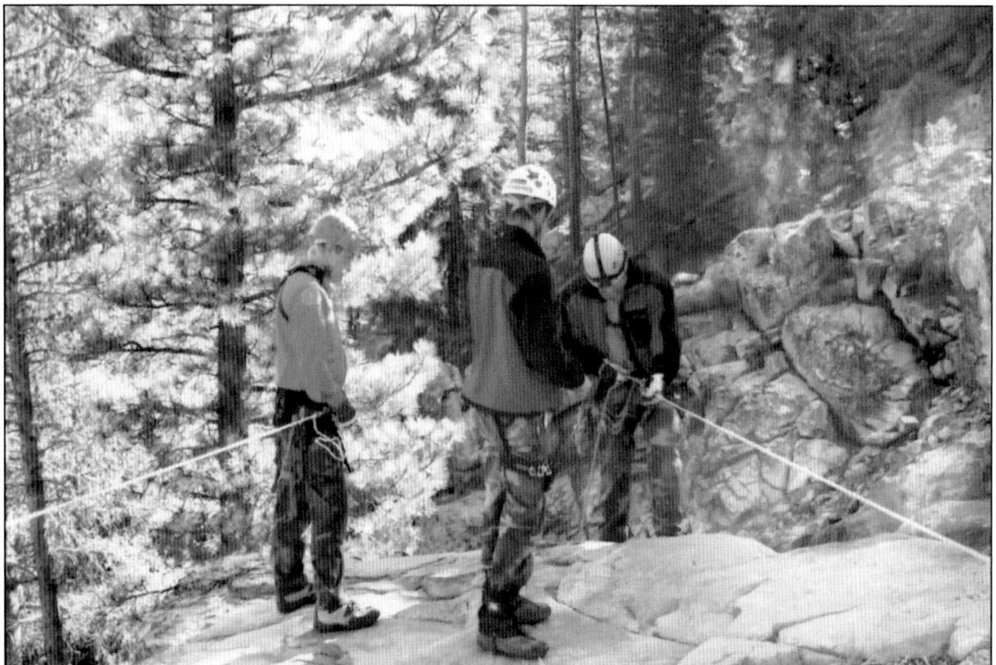

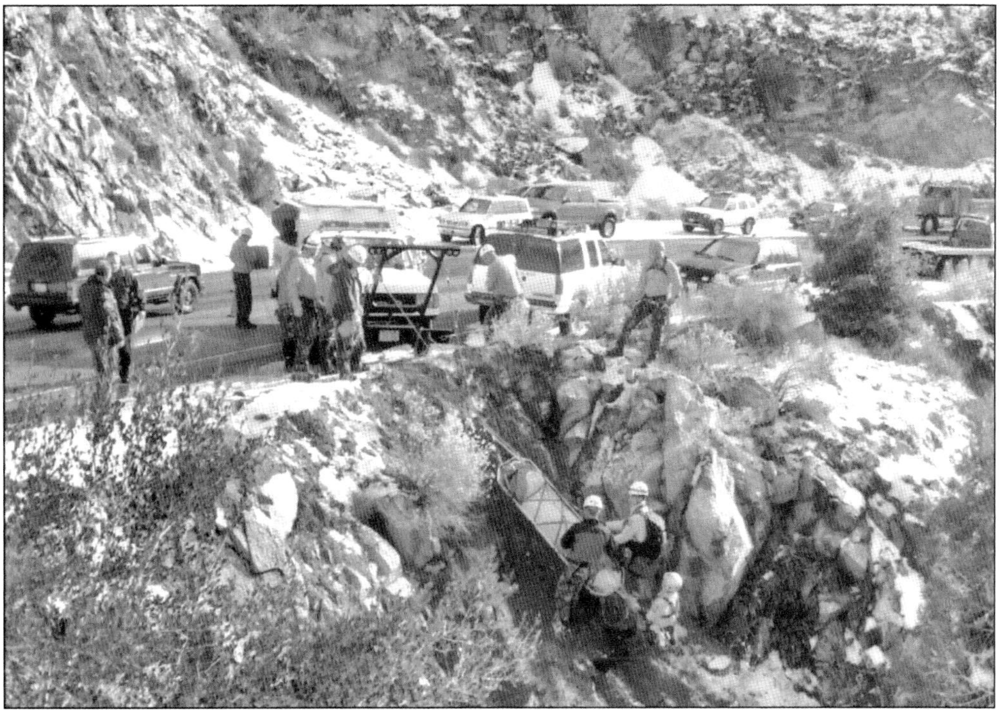

In December 2003, a woman was returning to her home in Angelus Oaks from Christmas shopping in Redlands when she hit a patch of ice and her car plummeted over the side of Highway 38. She did not survive. The team was called to recover her body and aid the tow truck in bringing up her car.

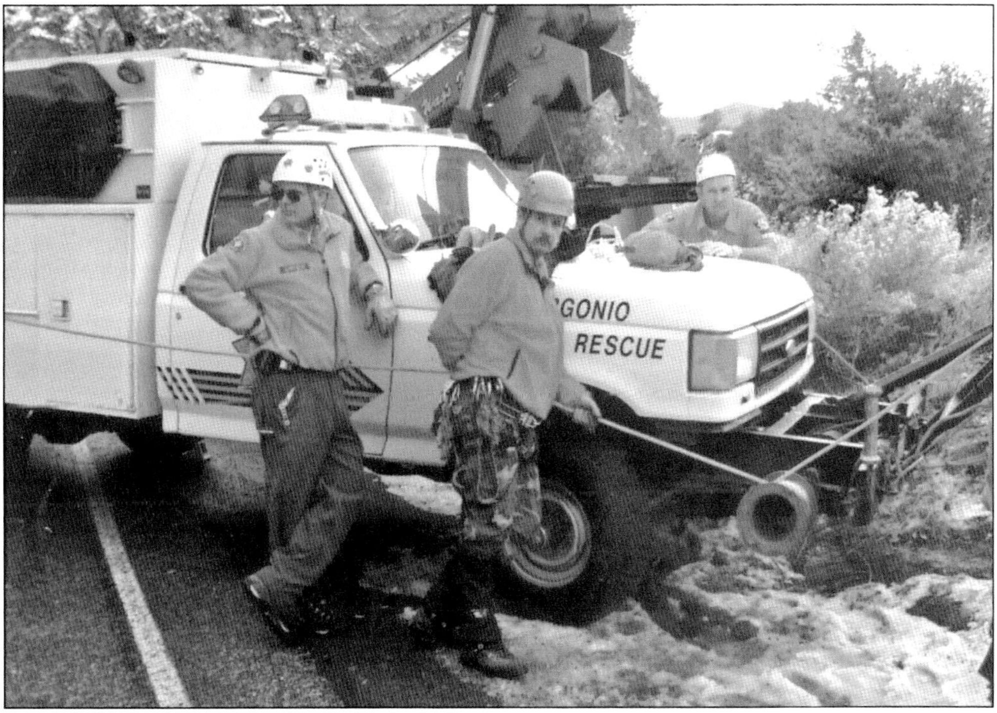

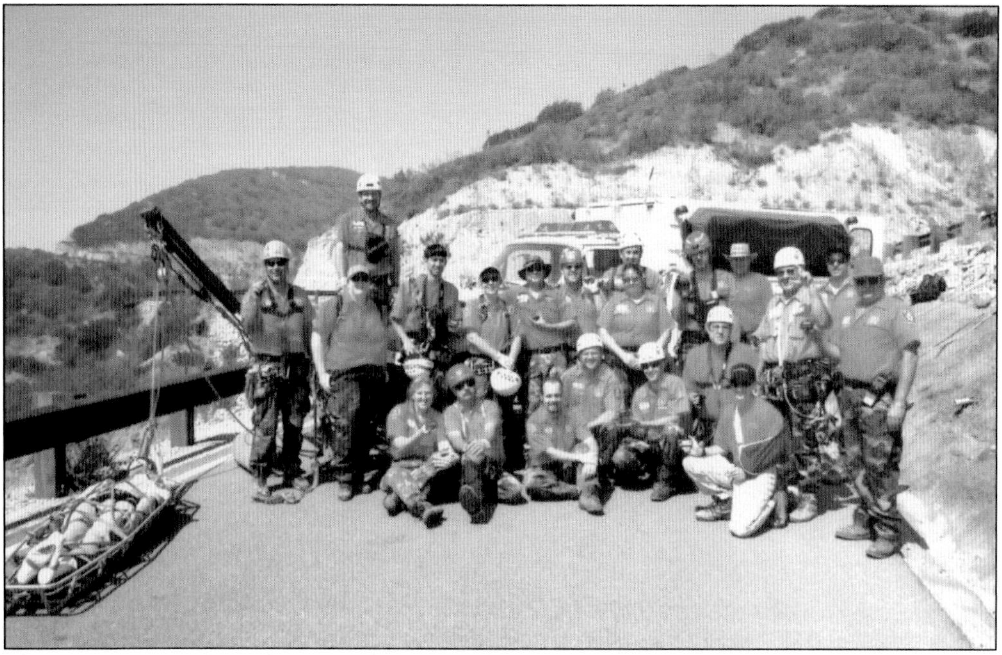

This group picture shows the team after successfully completing rope-rescue training on the Seven Oaks Dam in Highland, California. When no team members are willing to act as victims, the group also uses Rescue Randy, a full-size, 175-pound "dummy" who is probably one of the oldest active team members in the county.

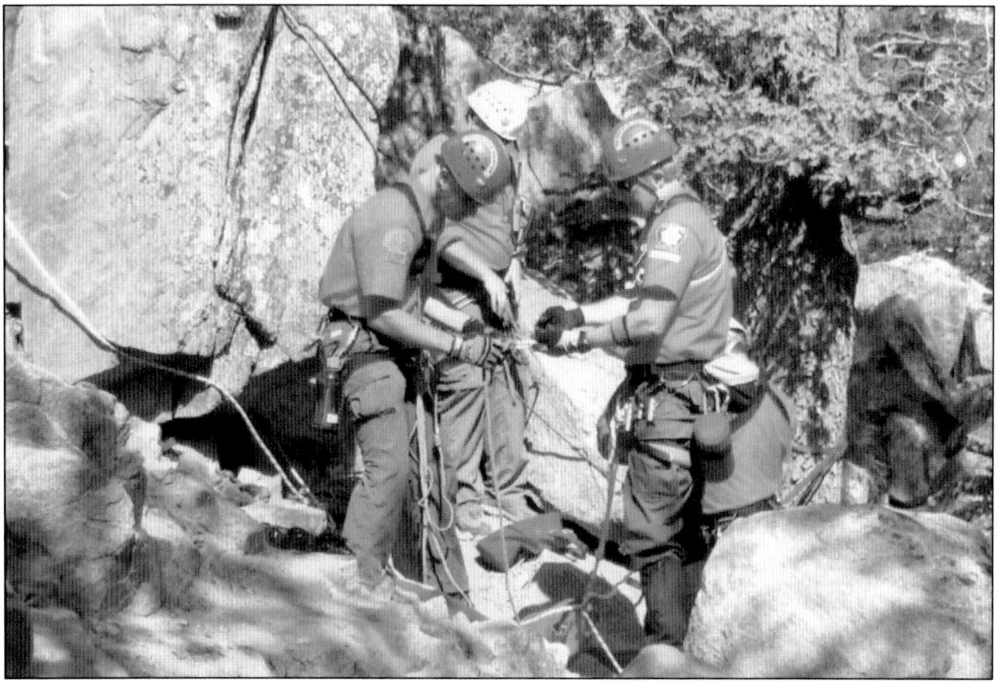

Here members tie a load-releasing hitch, one of the many knots used in a rope-rescue operation. The load-release hitch can be loosened to shift the load of the main line from one anchor to another, allowing team members to pass knots or fix equipment that is jammed.

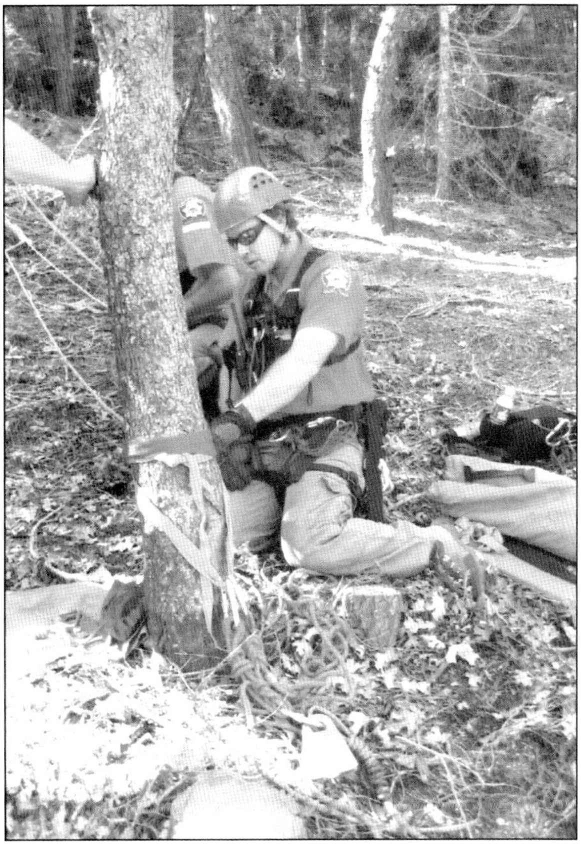

Bill Loenhorst, Ron Deitzman, and Dave Atchley tie webbing around trees to serve as anchors for the main line and the belay line for the upcoming rope-rescue operation. When one tree cannot be found that is large enough to be used as an anchor, team members will attach the anchor to trees or rocks behind it, using webbing or rope to make the anchor stronger. These attachments are called "back ties."

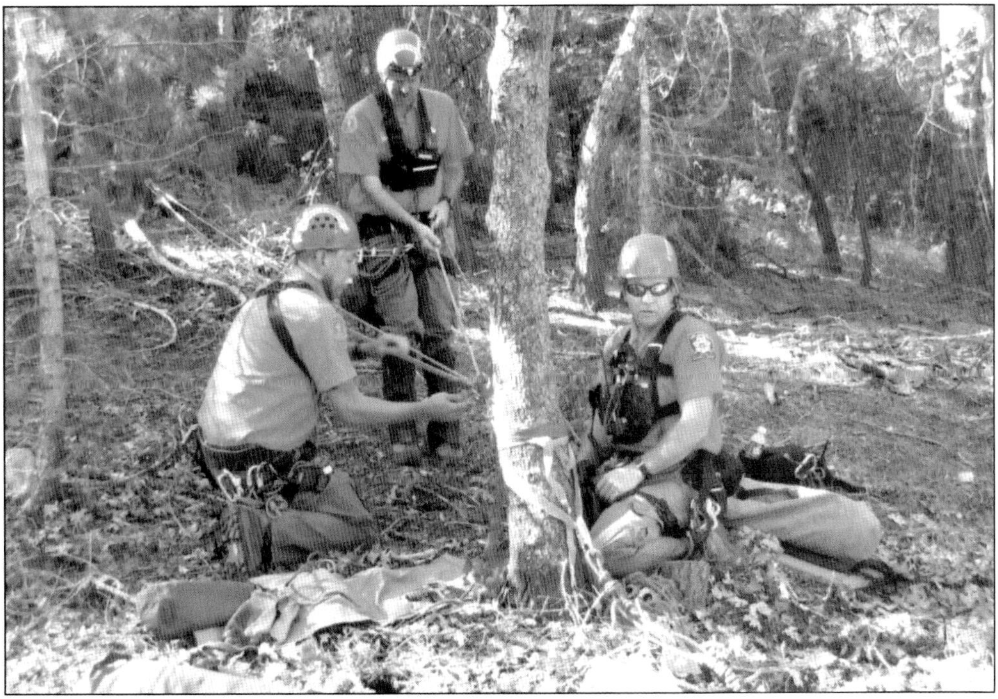

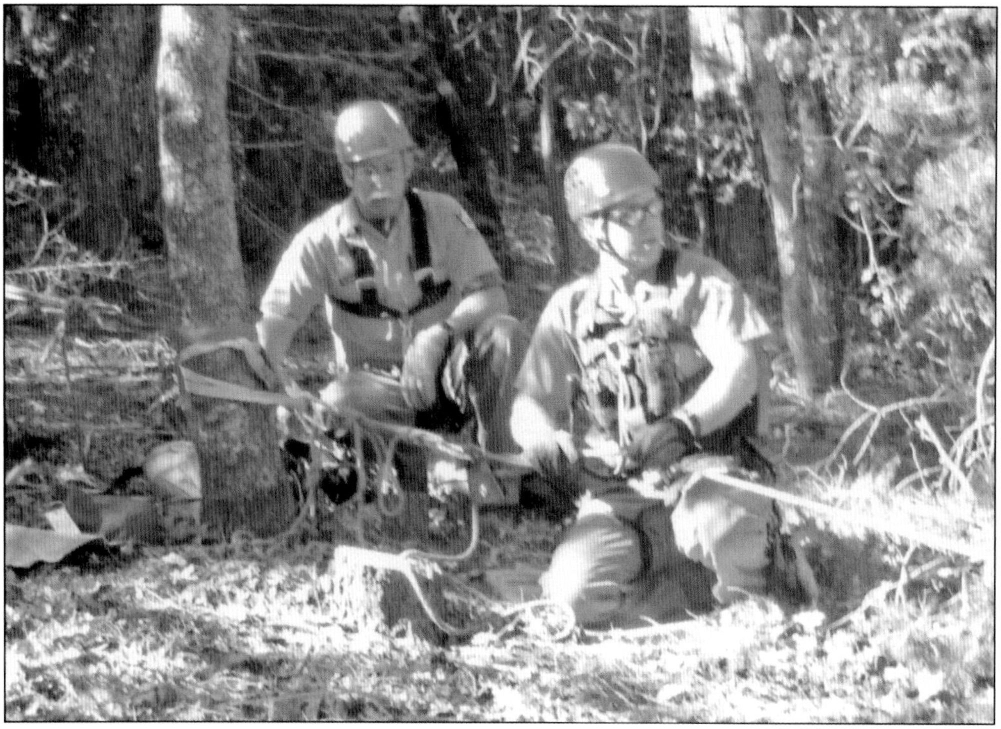

Ron Deitzman and Bill Loenhorst work the belay station during rope-rescue training in Angelus Oaks. The belay line is the safety line that holds the load if the main line fails during the raising or lowering operations of a rescue.

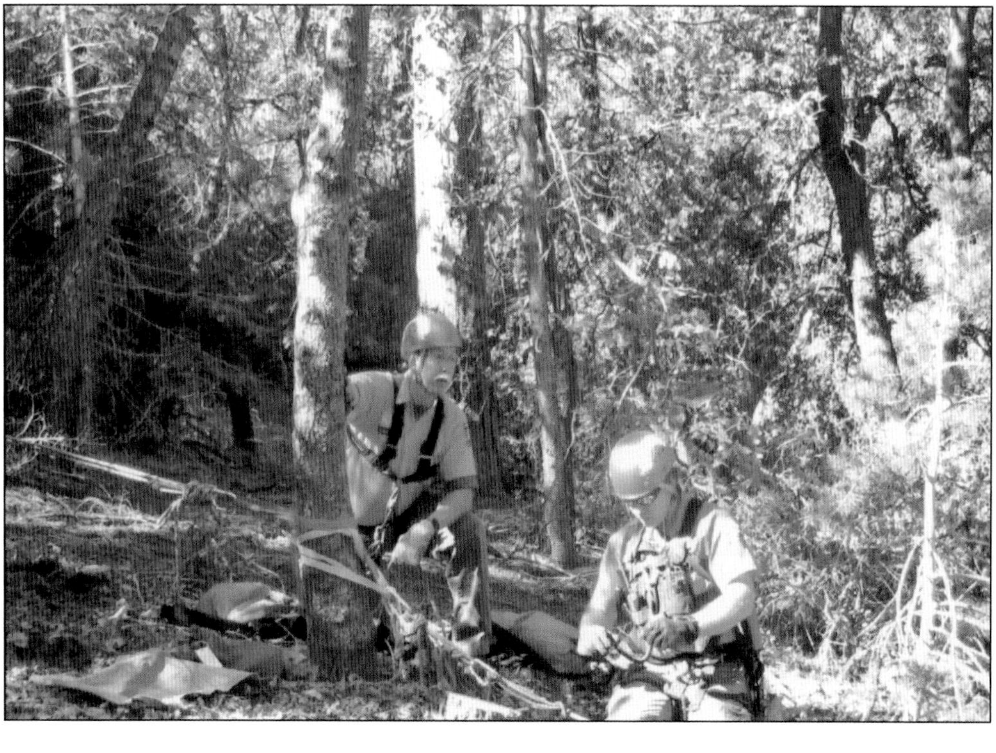

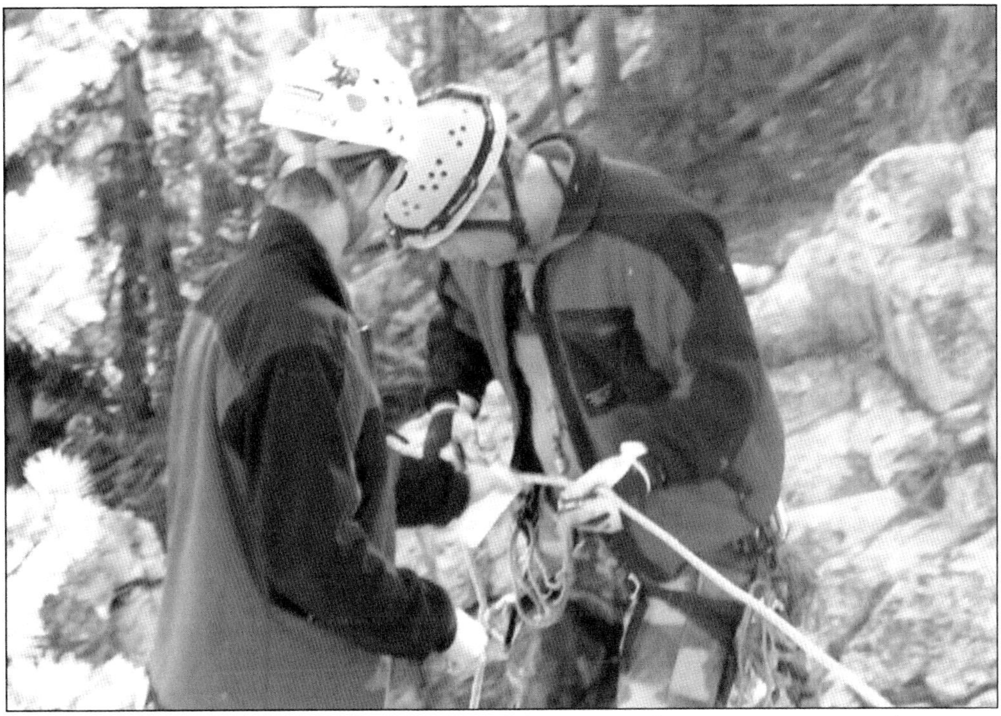

Scott Johnston, one of the team's rope-rescue experts, is teaching newer team members rope-rescue knots and the proper use of rescue equipment. The team's success in rope-rescue operations depends on people like Scott, who take advanced rope-rescue classes and return to teach the latest in rope techniques.

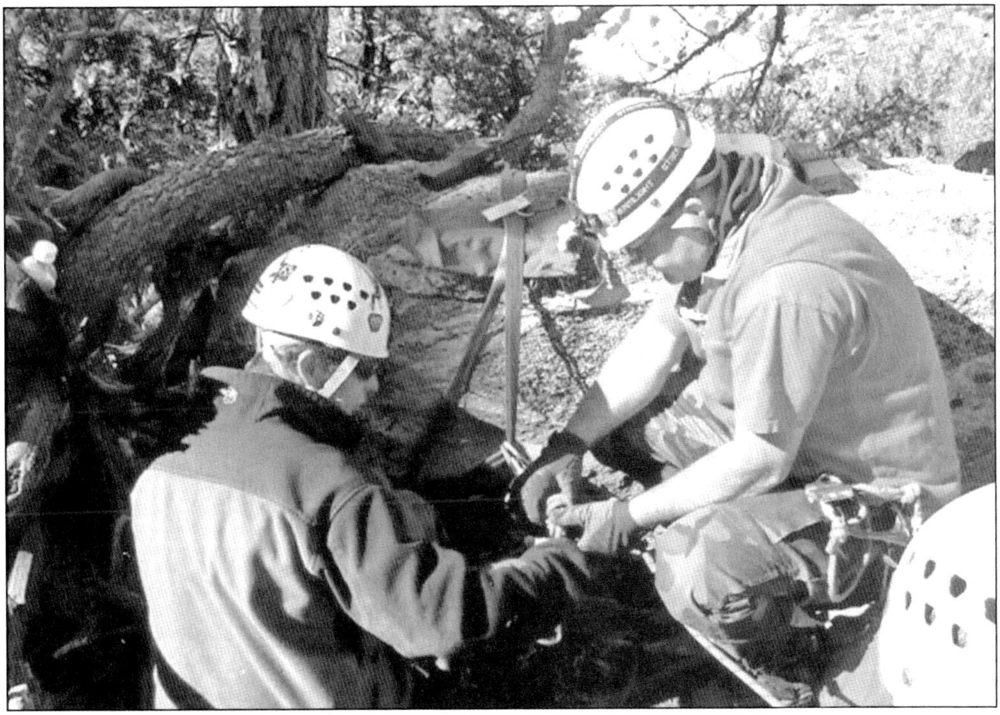

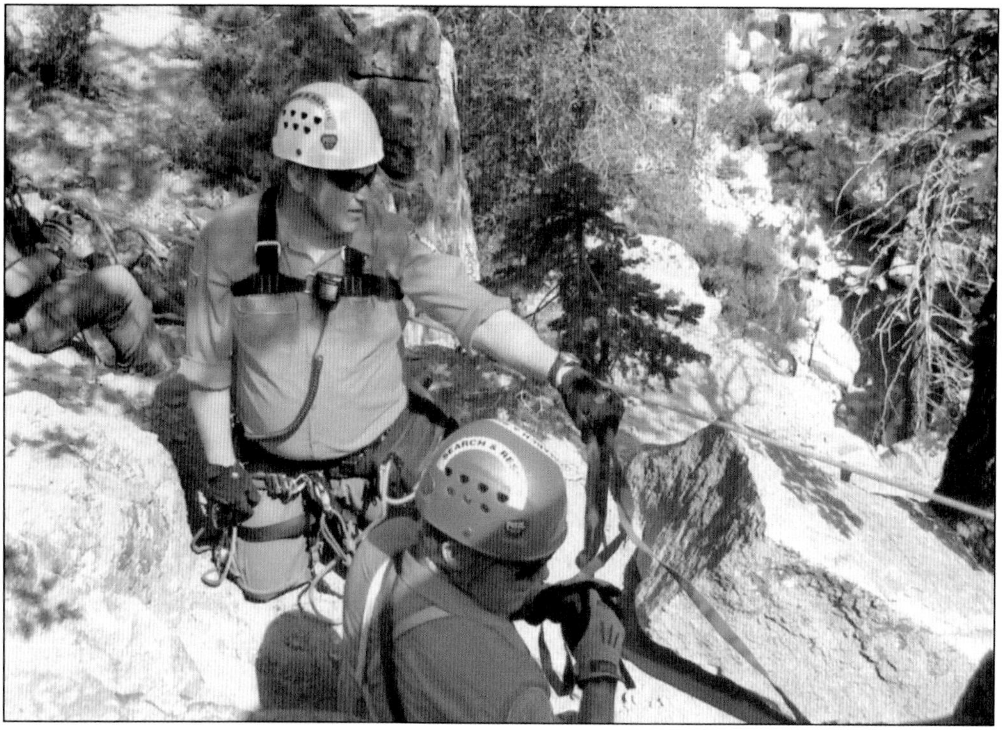

Shane Elliott works to set up the main line during team training at Deep Creek near Running Springs, California. The main line is the rope responsible for raising and lowering the litter and the litter attendant.

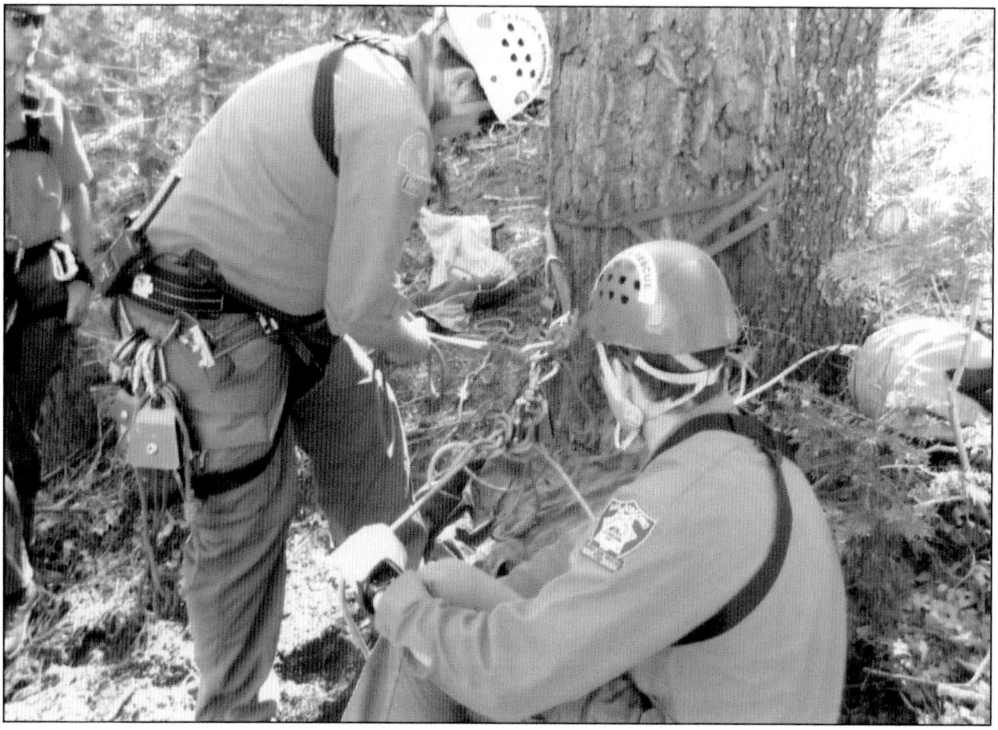

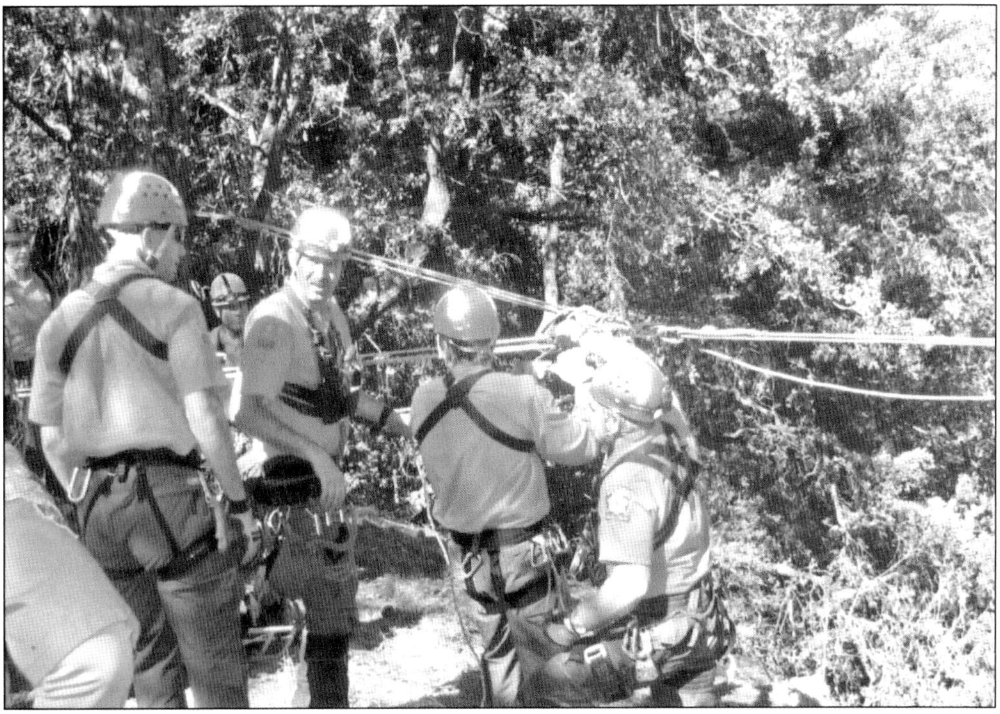

Dave Atchley acts as safety officer during this rope-rescue training session near Angelus Oaks, California. The safety officer checks all of the rigging before the rope-rescue operation begins.

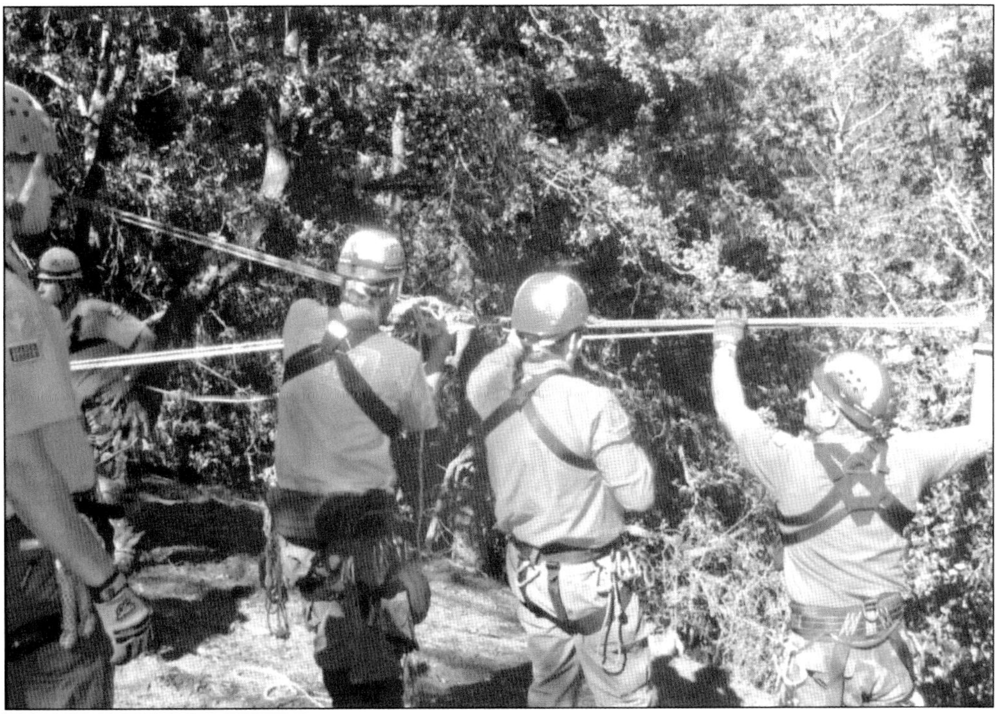

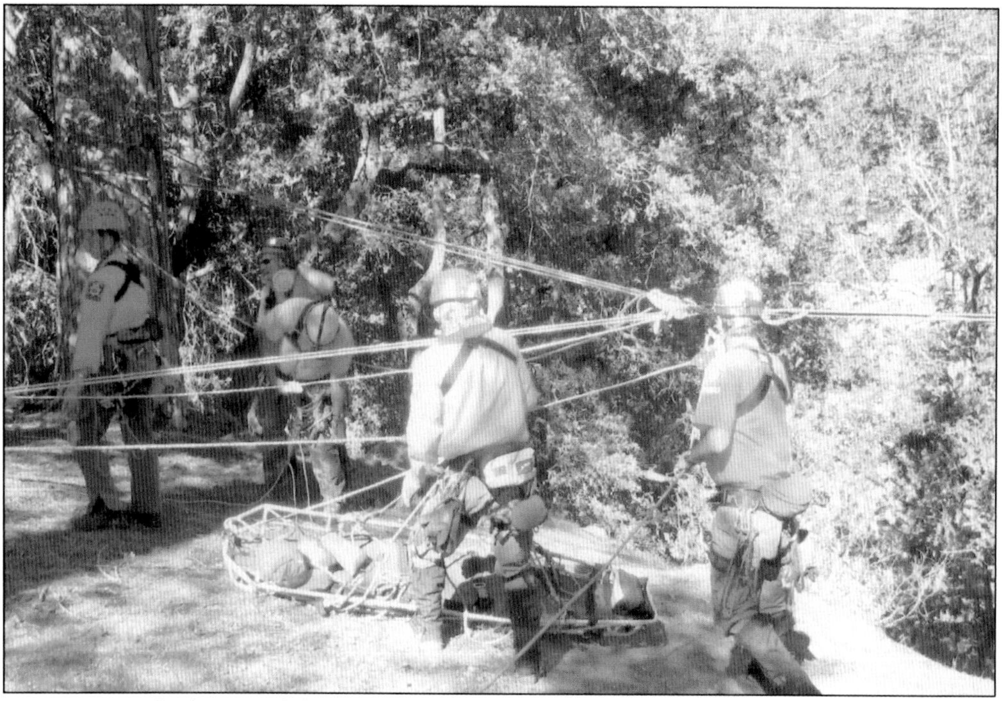

To maneuver the litter and the litter attendant over a difficult edge, the team rigs a high focal point by tying ropes between trees near the edge to bring the main line up off the ground. The main line rope is put into a pulley that hangs from the high focal point. The higher the focal point and the pulley, which holds the main line, the easier it is for the litter and litter attendant to bring the victim up the steep side of a rock face and over the edge of the cliff to safety.

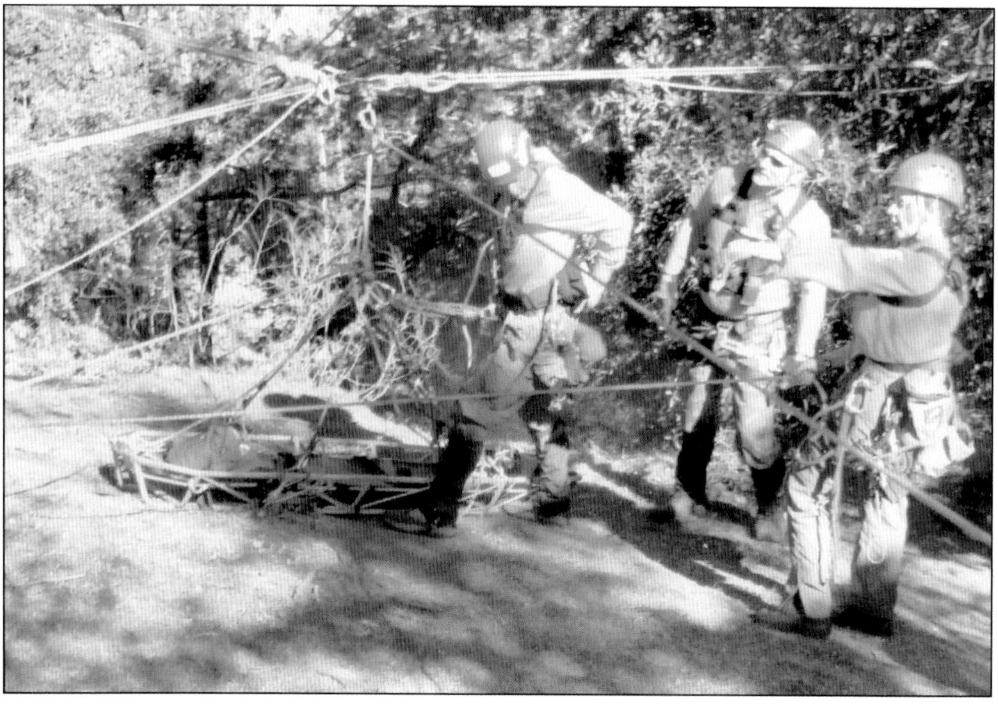

Five

HELITAC

Since the beginning of the San Gorgonio Search and Rescue, helicopters have been used to aid in rescue operations. Helicopters are able to insert team members into the wilderness and save hours of hiking to reach victims. They can also be used to extract injured hikers and quickly transport them to medical aid. The team regularly trains with Sheriff's Aviation and has a good working relationship with all of the pilots and crew members. It is only fitting, and a point of team pride, that when a sheriff's helicopter crashed in the mountains, it was the San Gorgonio team that located the downed helicopter and helped rescue of the crew.

Around 5:00 in the evening on March 9, 1999, a San Bernardino County sheriff's helicopter crashed when a cloud suddenly engulfed it and the pilot lost his bearings. All of the volunteer rescue teams in the county were immediately activated. Around 9:00 p.m., the incident commander sent the team to take Emergency Locater Transponder (ELT) readings on top of Monument Peak. They got an extremely strong signal coming from directly below them in a small, steep canyon. The team began to slowly descend into the canyon, which was filled with thick manzanita. Around 3:00 a.m., the team popped out of the brush and could see a lonely flashlight waving at them. At 5:00 a.m., they broke through a wall of branches and there was the helicopter lying on its side, the pilot and crew huddling together for warmth. At 7:00 a.m., the sheriff's department rescue helicopter arrived; they hoisted the crew out of the crash site and into the rescue ship and took them to the hospital for further examination. Daylight revealed that the helicopter was perched sideways on a small flat area just above a vertical cliff. If the helicopter had crashed a few yards to either side, they would have fallen to the bottom of the canyon to their deaths. The rescue ship returned to pick up the rescue team members.

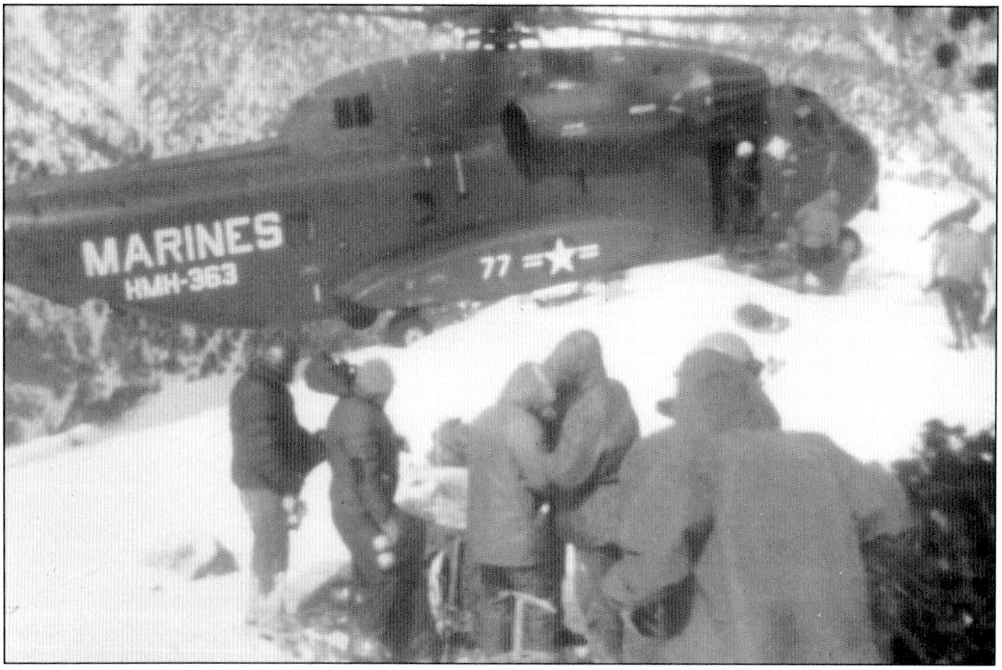

A plane crashed on Cucamonga Peak in 1959 killing all seven passengers and the pilot. The marines were called to aid in the recovery of the bodies, transporting team members to the crash site and later bringing everyone back to the command center.

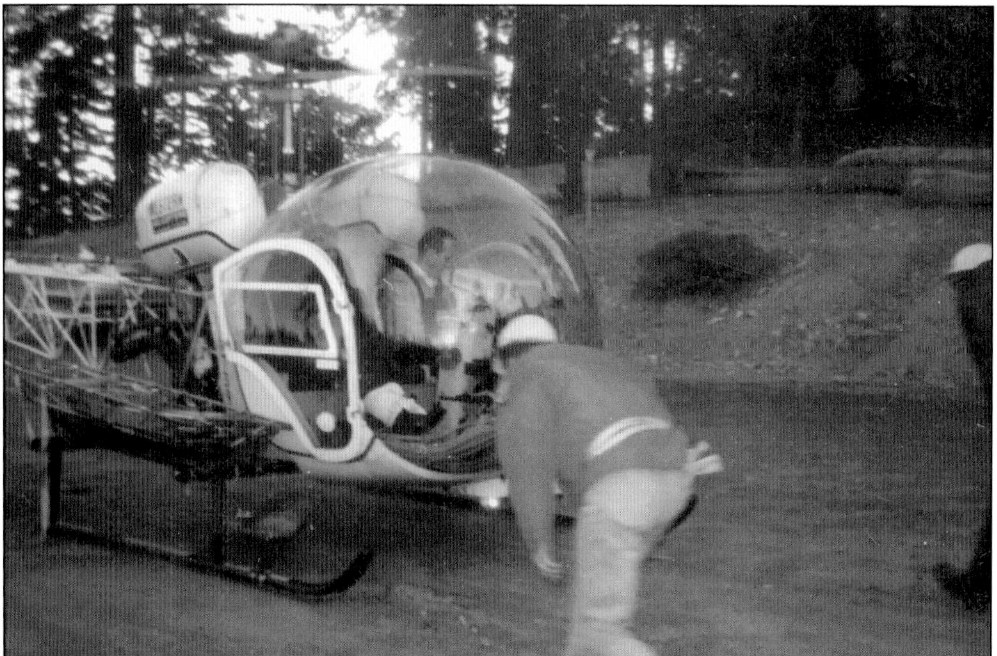

The first helicopters the sheriff's department used could only hold one passenger at a time. Even though it took many trips to insert and extract team members from the wilderness, the helicopter still saved valuable time for the search-and-rescue group.

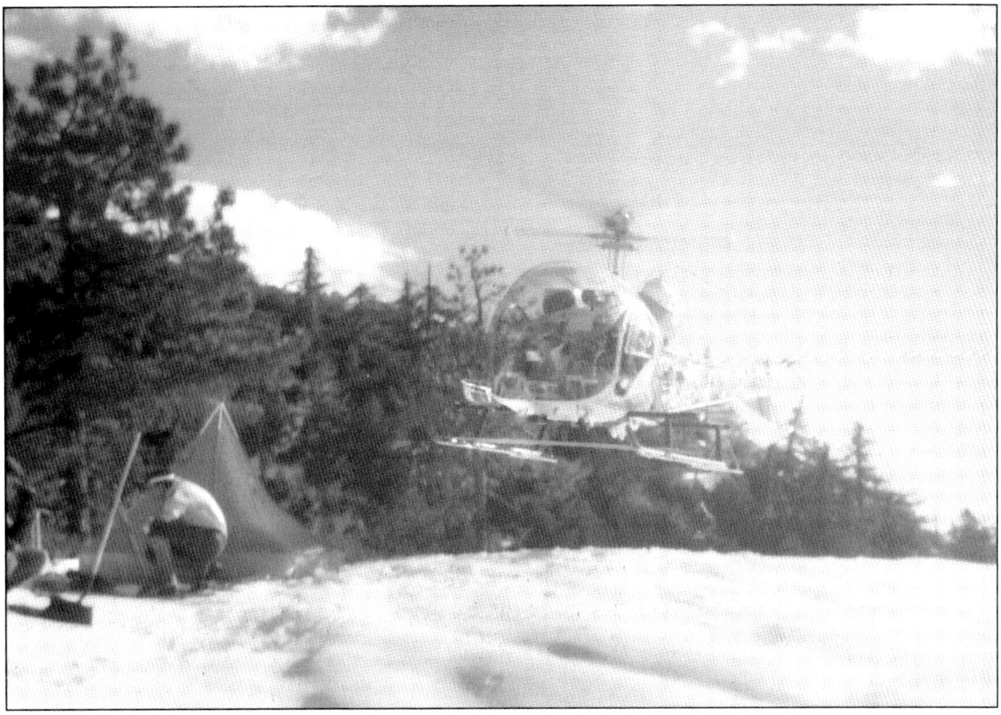

A helicopter lands on the snow near San Bernardino Peak to pick up two members who have been searching for a lost hiker. They spent the night in a shelter after searching the previous day.

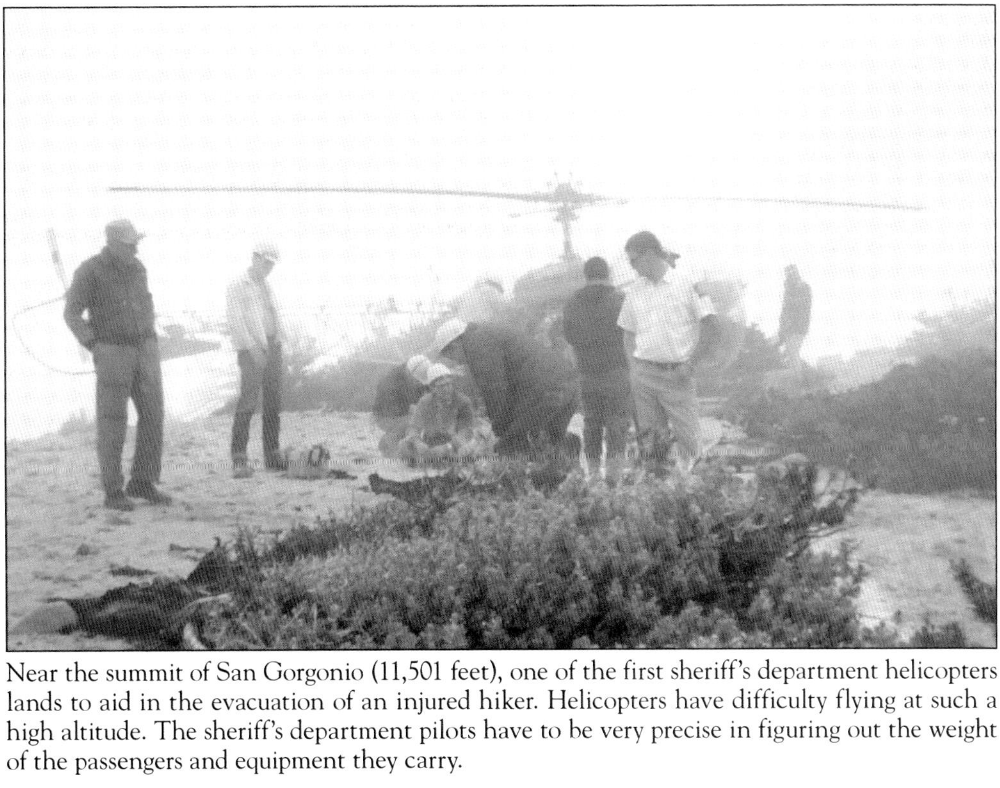

Near the summit of San Gorgonio (11,501 feet), one of the first sheriff's department helicopters lands to aid in the evacuation of an injured hiker. Helicopters have difficulty flying at such a high altitude. The sheriff's department pilots have to be very precise in figuring out the weight of the passengers and equipment they carry.

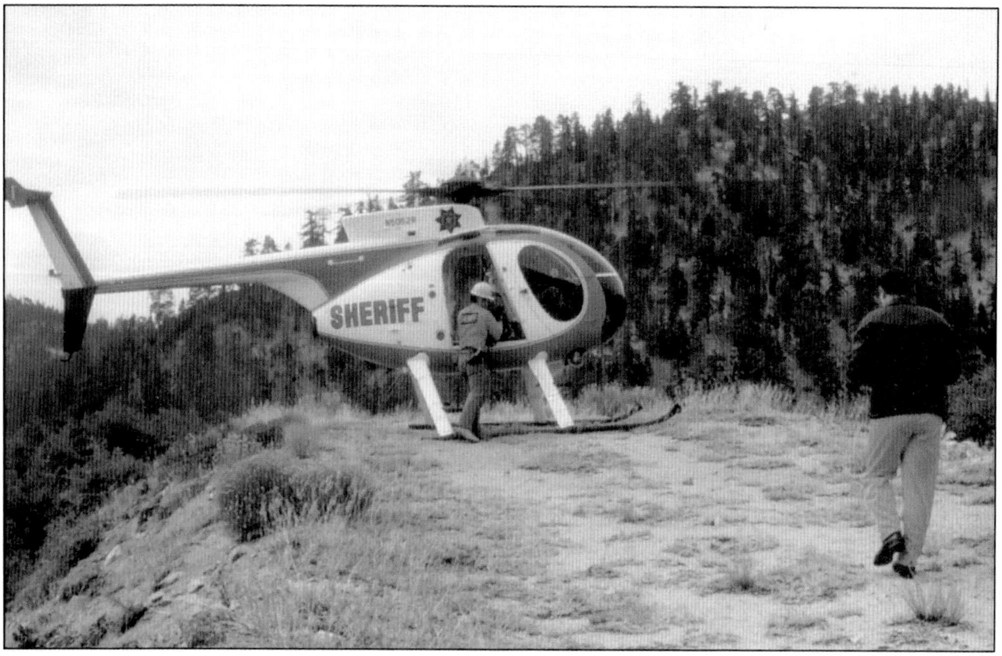

Sheriff's Aviation drops off team members on the east side of the Mill Creek Jump Off to recover the bodies of two people from a plane crash. The Mill Creek Canyon is a box canyon, meaning the two ridge lines that form the canyon come together, making a wall of loose rock at the end of the canyon. This wall is called Mill Creek Jump Off.

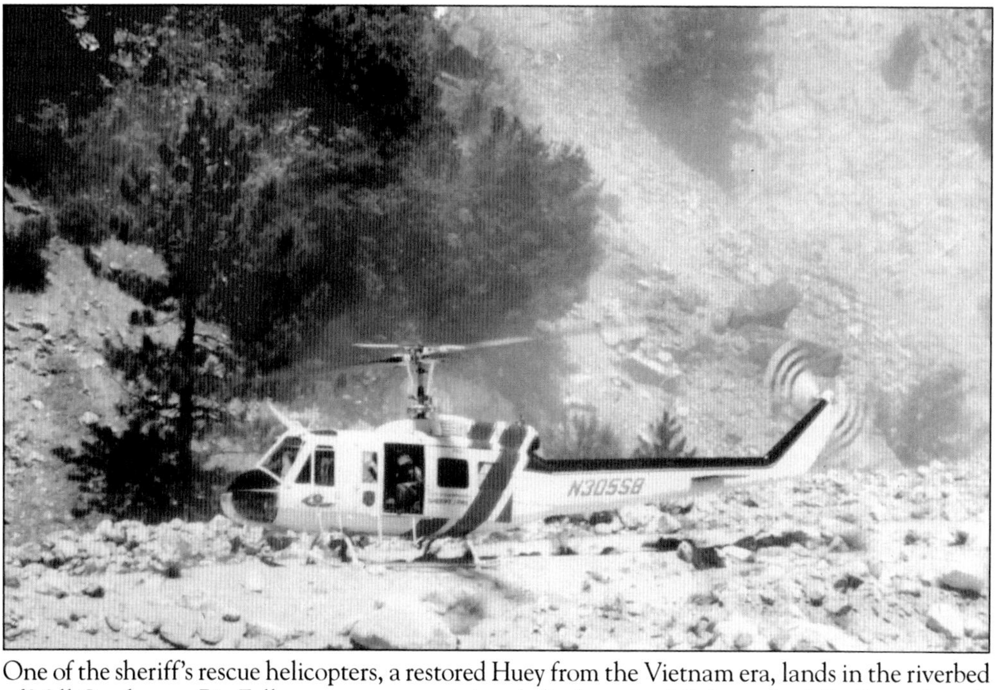

One of the sheriff's rescue helicopters, a restored Huey from the Vietnam era, lands in the riverbed of Mill Creek near Big Falls to evacuate an injured climber who fell from the falls. Even though signs have been posted by the U.S. Forest Service prohibiting climbing of the falls, every year hundreds of people attempt to climb them, many are injured, and a few lose their lives.

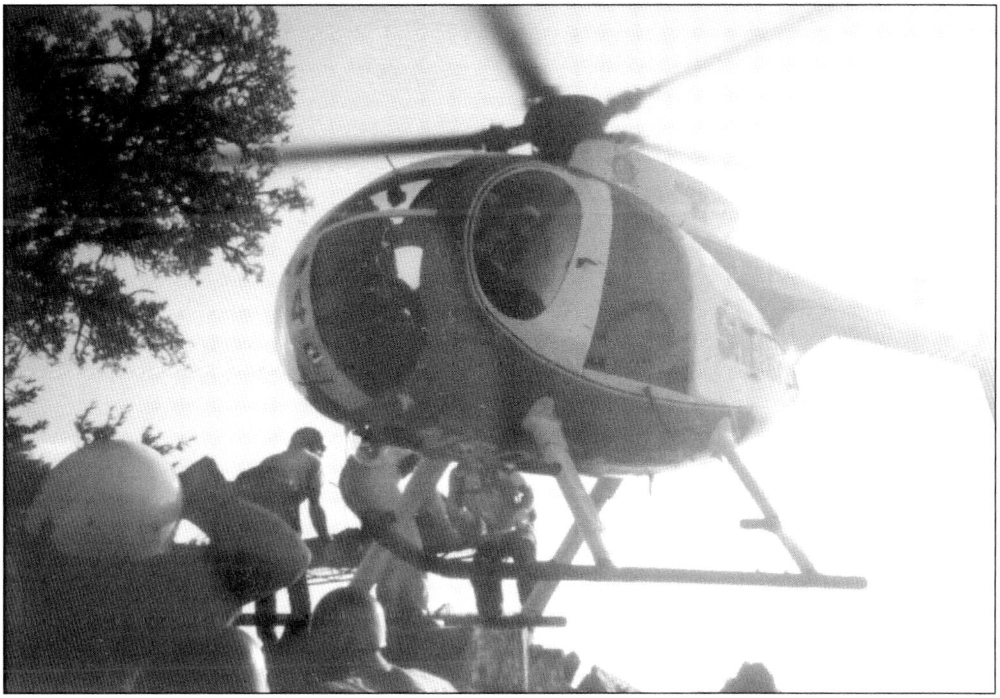

When a hiker fell near the palisades below San Bernardino Peak, Sheriff's Aviation once again aided team members in rescuing the injured person. The pilot was only able to set down one skid on the small outcropping of rocks. Team members had to climb out on the skid suspended 2,000 feet above the valley floor to enter the helicopter.

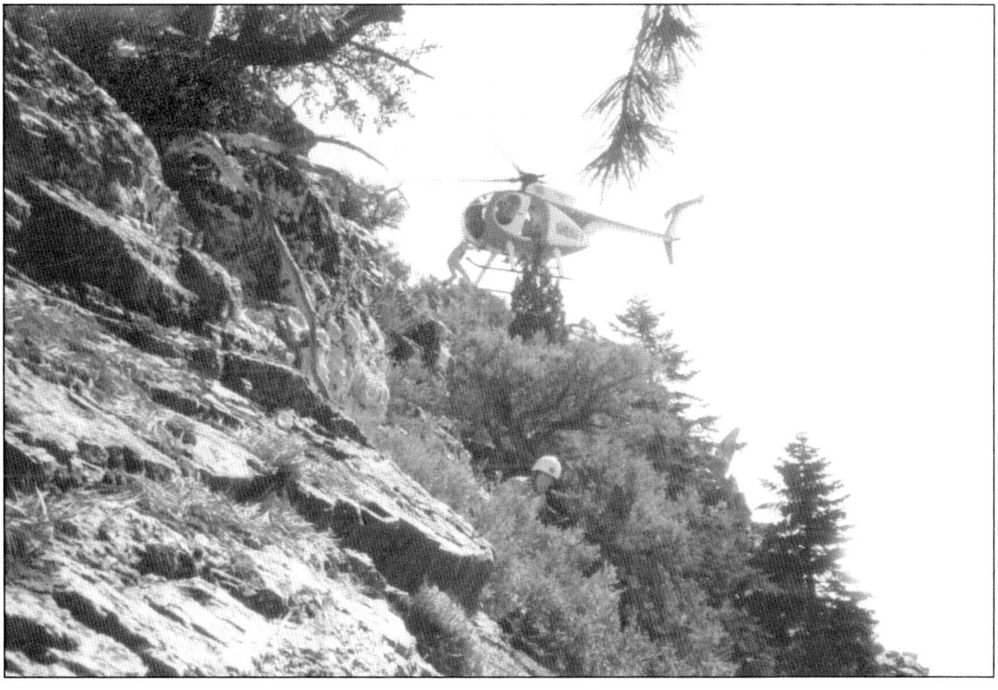

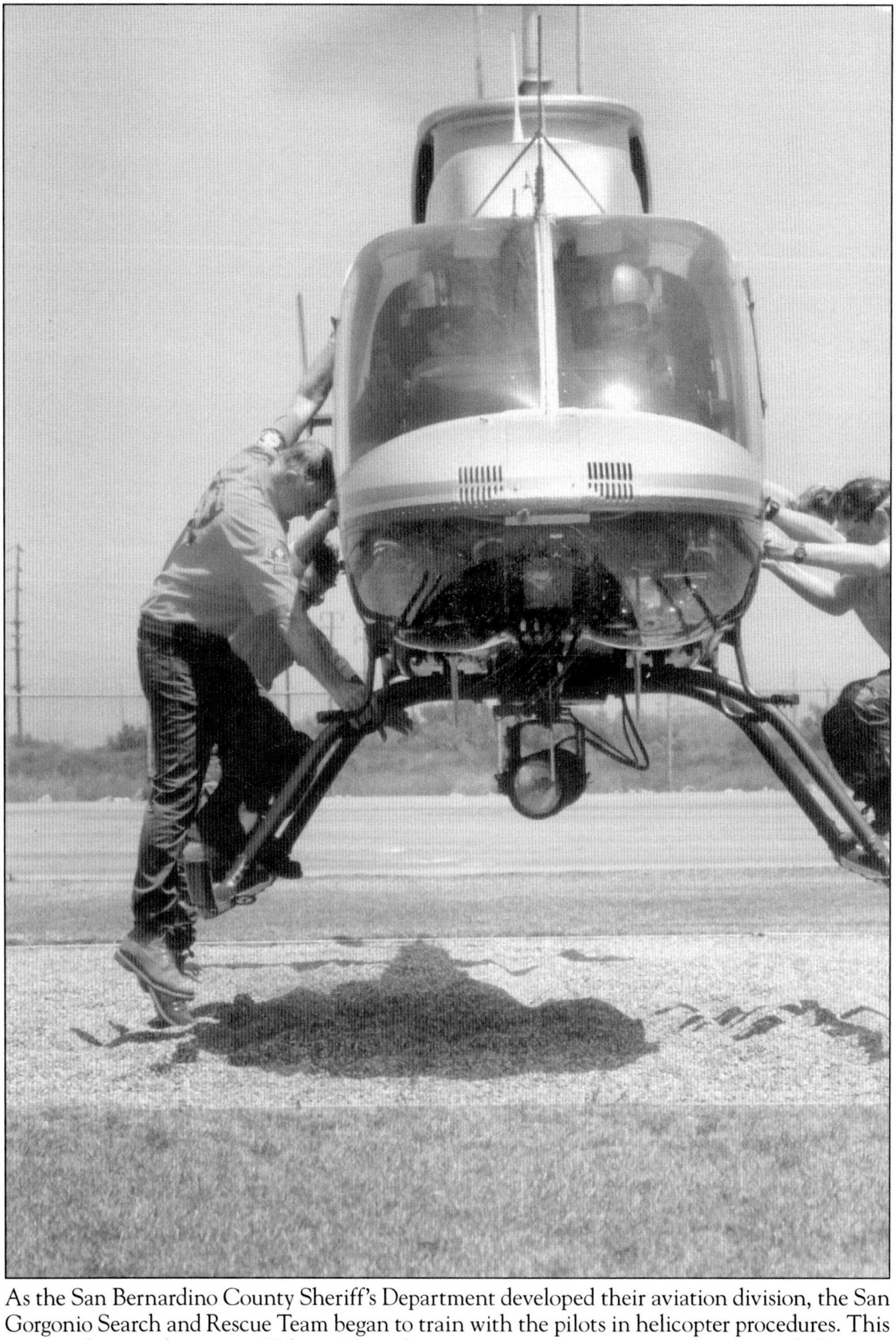

As the San Bernardino County Sheriff's Department developed their aviation division, the San Gorgonio Search and Rescue Team began to train with the pilots in helicopter procedures. This training became known as Helitac. Here the team practices entering and exiting a helicopter while the airship hovers several feet off the ground.

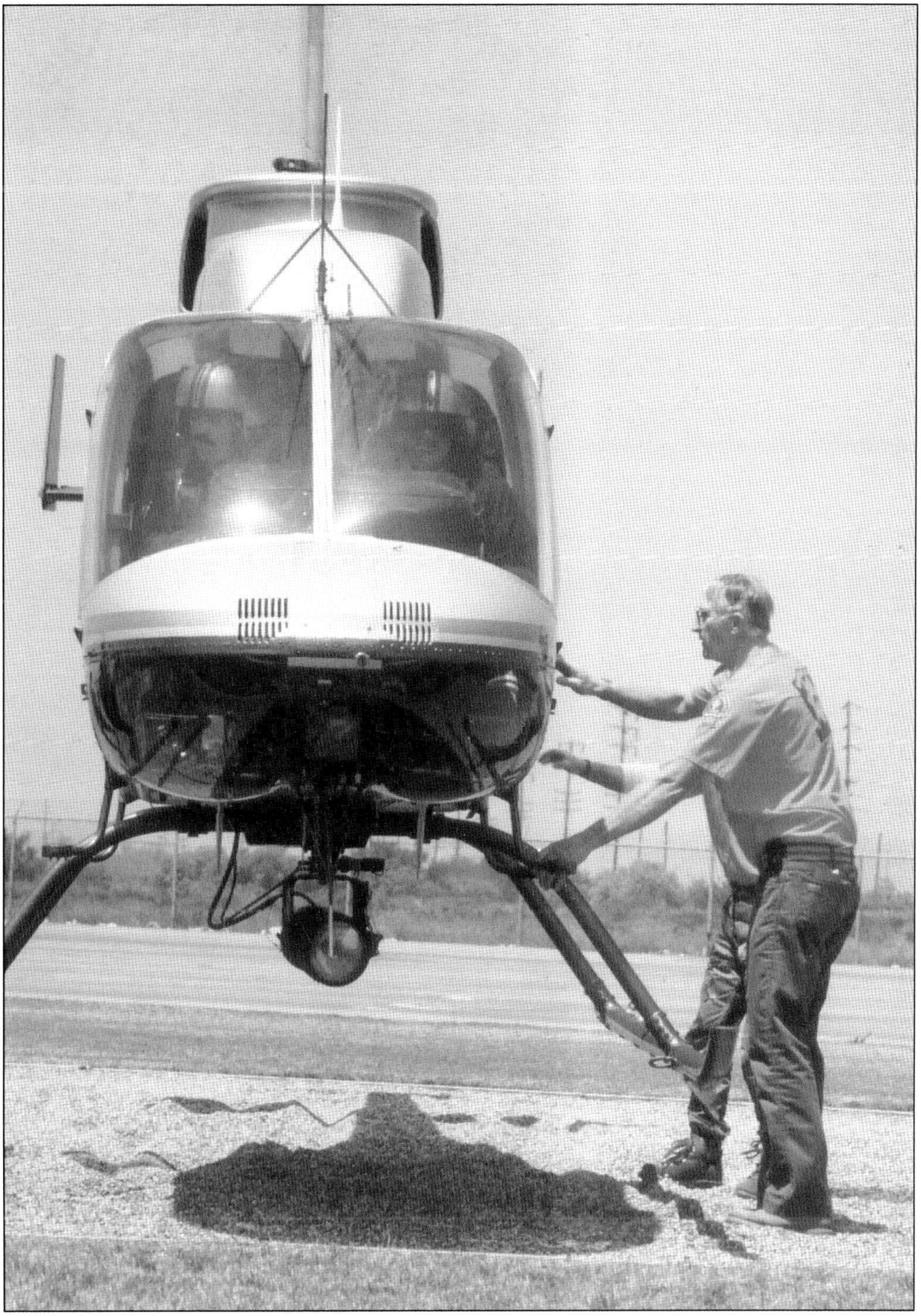

Team member Patrick "Packy" Wassell performs the maneuver called a hover-step, which has been used numerous times in actual wilderness rescues. Many times the terrain is not level enough for the helicopter to land safely, so it will hover off the ground while team members load injured people into the helicopter.

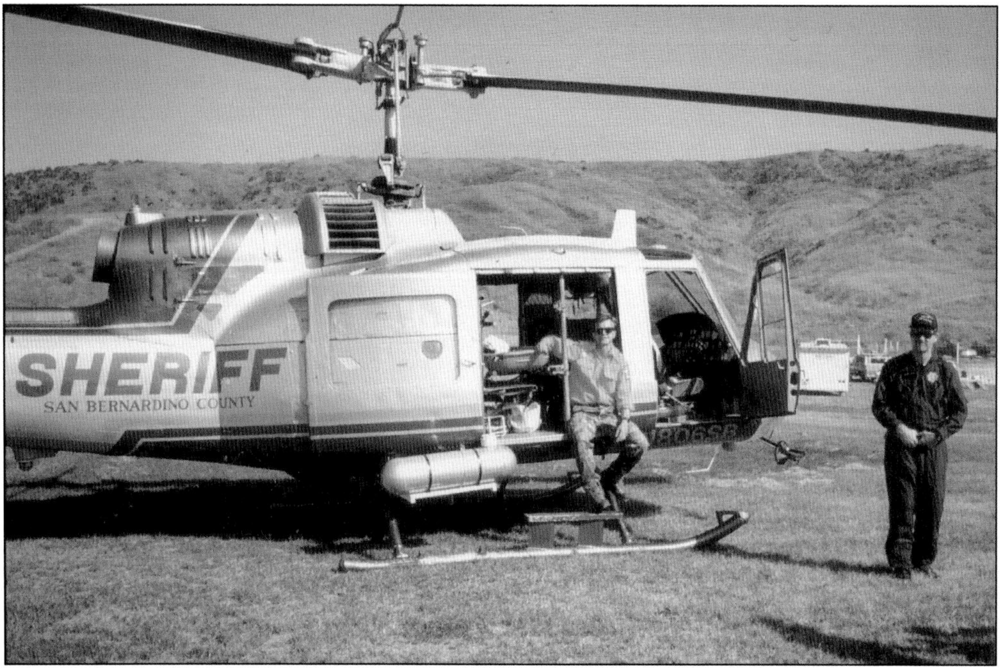

In the late 1970s, the sheriff's department acquired a Huey helicopter, probably through army surplus from the Vietnam War. The Huey was a valuable asset to the San Gorgonio Search and Rescue Team. Its strong engine is ideal for high altitude flying and can carry four rescuers and equipment in one trip.

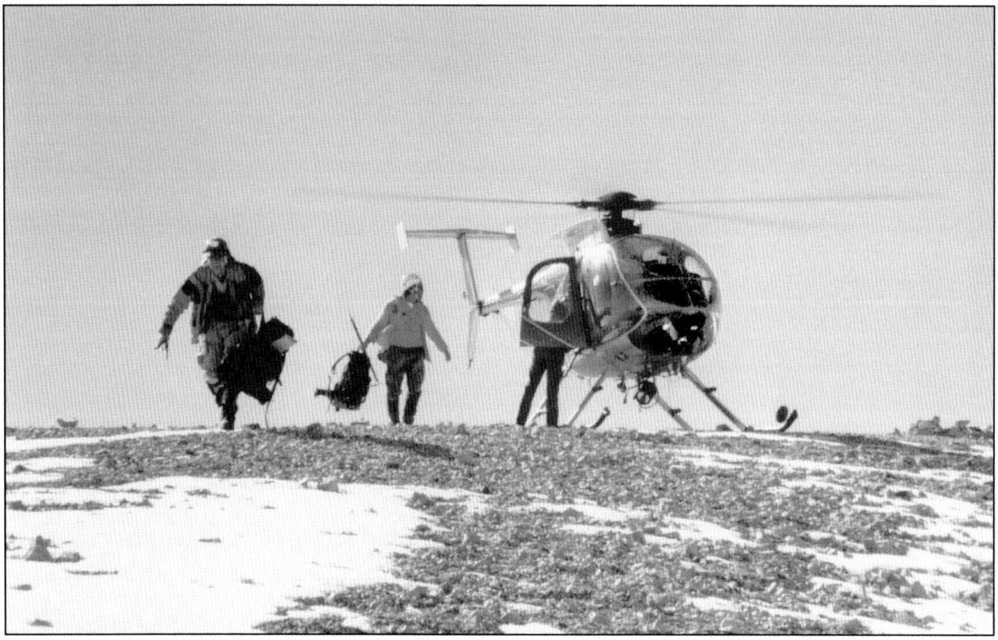

During the summer of 1995, Frank Ruiz and Wayne Daniels were deposited on the summit of Mount Baldy to search for a lost hiker. Mount Baldy, also known as Mount San Antonio, is the tallest mountain—at 10,064 feet—in the San Gabriel Mountains, just north of Rancho Cucamonga, California. Its rugged terrain has claimed the lives of many hikers over the years.

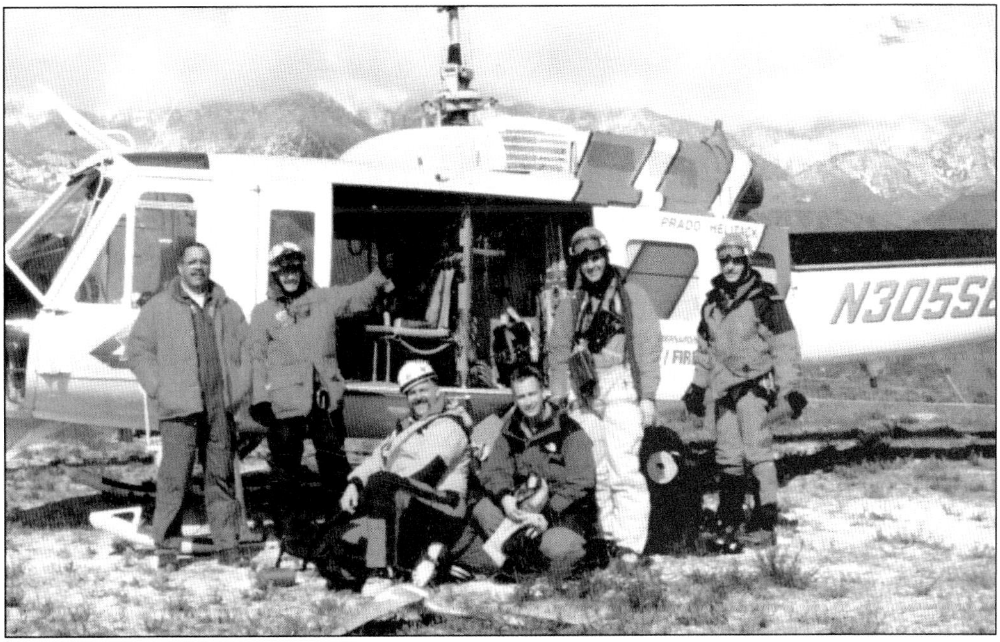

In February 1998, a plane taking off from Big Bear Airport crashed into the east side of Mount San Gorgonio in the middle of a winter storm. After the storm lifted, San Gorgonio team member Rod Walters, along with rescuers from West Valley and Central's rescue teams, flew to area where the plane crashed. Unfortunately, the snow was too deep to recover the bodies at that time, and the recovery was made in June when the snow had melted.

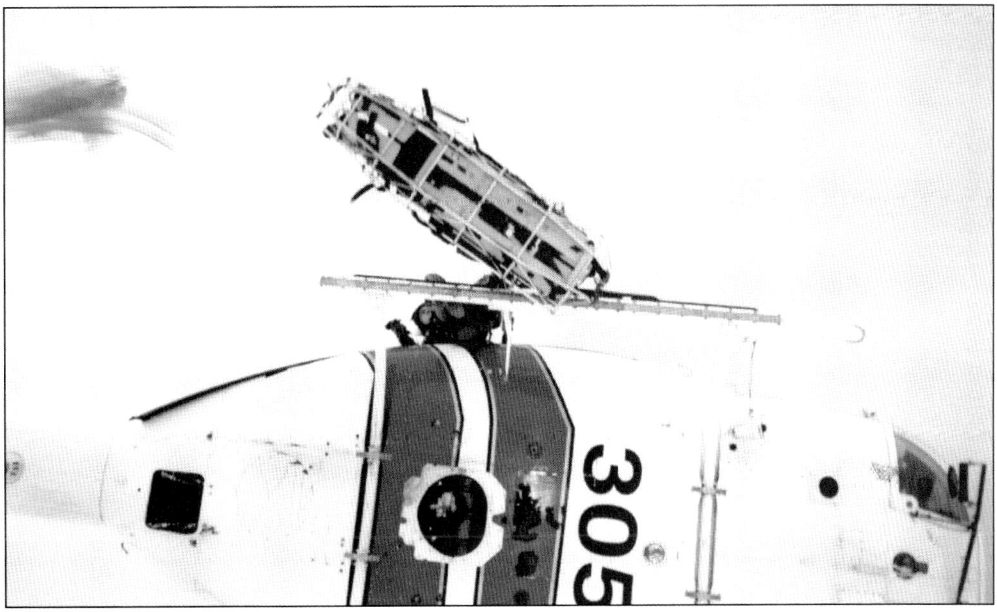

In January 2005, San Gorgonio team member Rod Walters and Valley of the Falls team member Dr. Jim Evans were flown into the Mill Creek Jump Off to rescue an injured ice climber. Unfortunately, the man had died of his injuries before rescuers could reach him. Equipping the rescue Hueys with hoists gives the sheriff's department the ability to pluck injured and deceased hikers and climbers out of the wilderness.

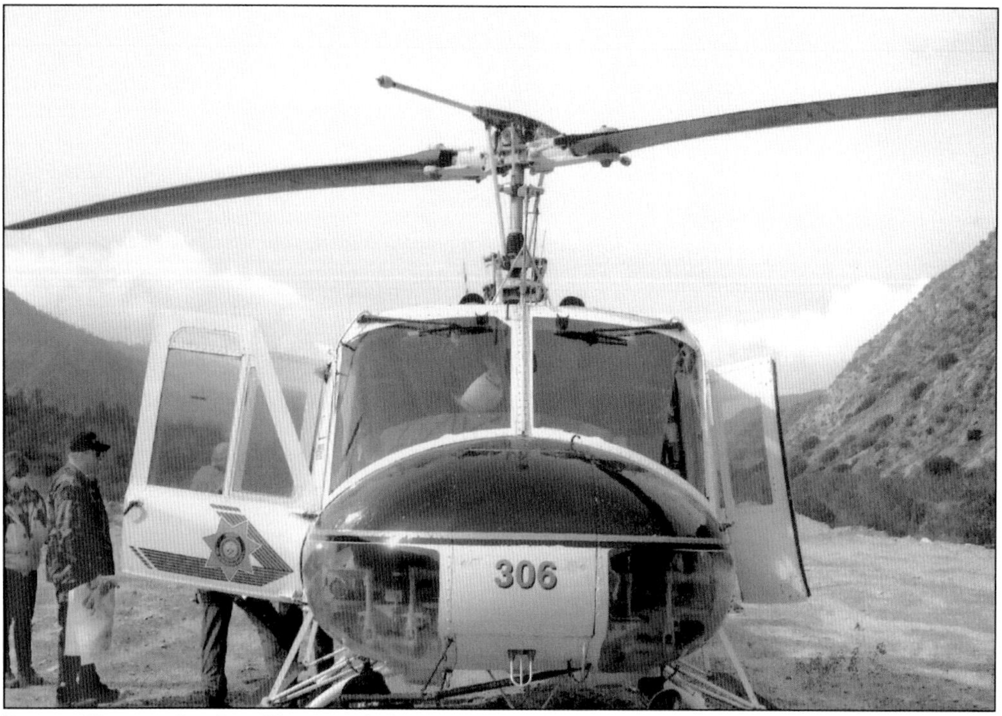

Pictured here is the sheriff's rescue helicopter parked at the landing zone in Mill Creek Canyon. The pilot and team members discuss the best way to retrieve an injured hiker from the snowy peak of San Gorgonio. In January 2004, the San Gorgonio Wilderness received an extraordinarily high volume of snow, causing very dangerous conditions in the mountains.

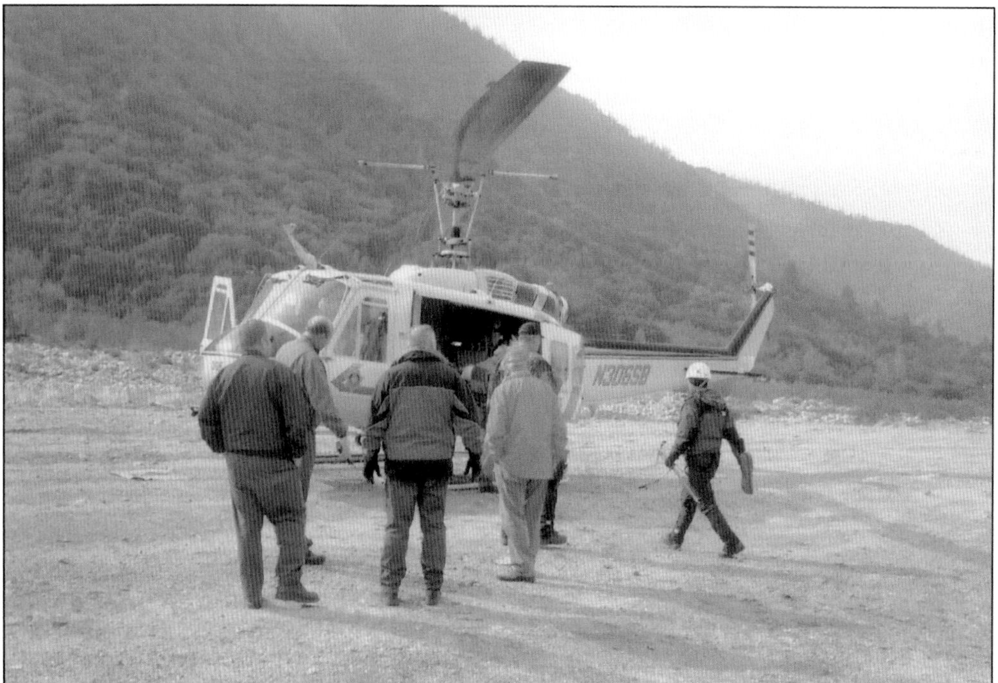

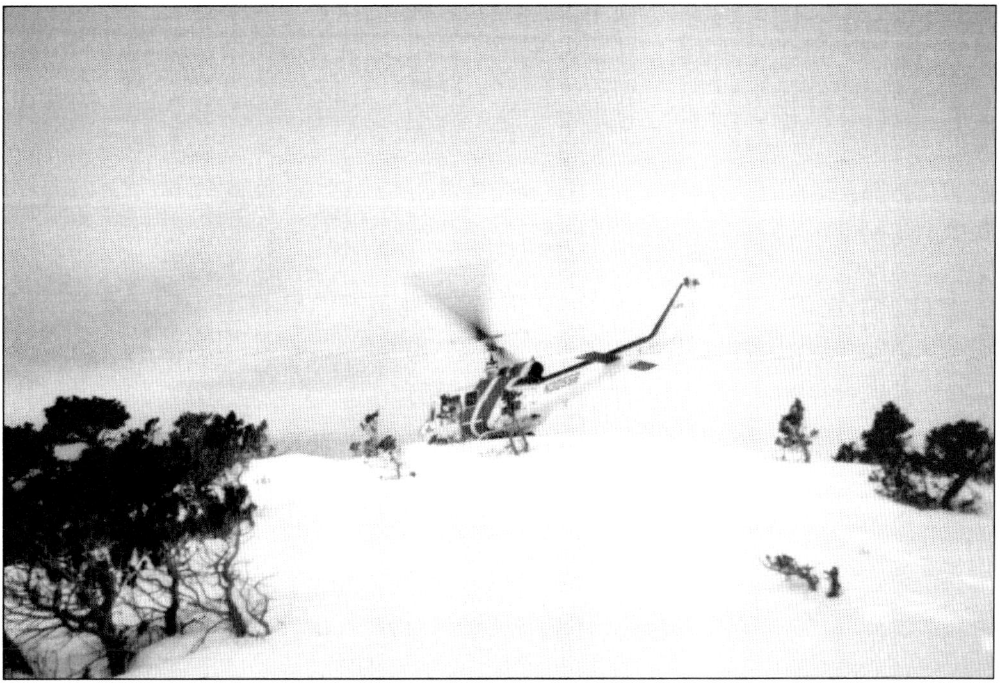

The sheriff's helicopter lands near the summit of San Gorgonio to drop off team members to search for lost hikers in the winter of 2004. The group would make its way down the Vivian Creek Trail searching for an overdue hiker out of San Diego.

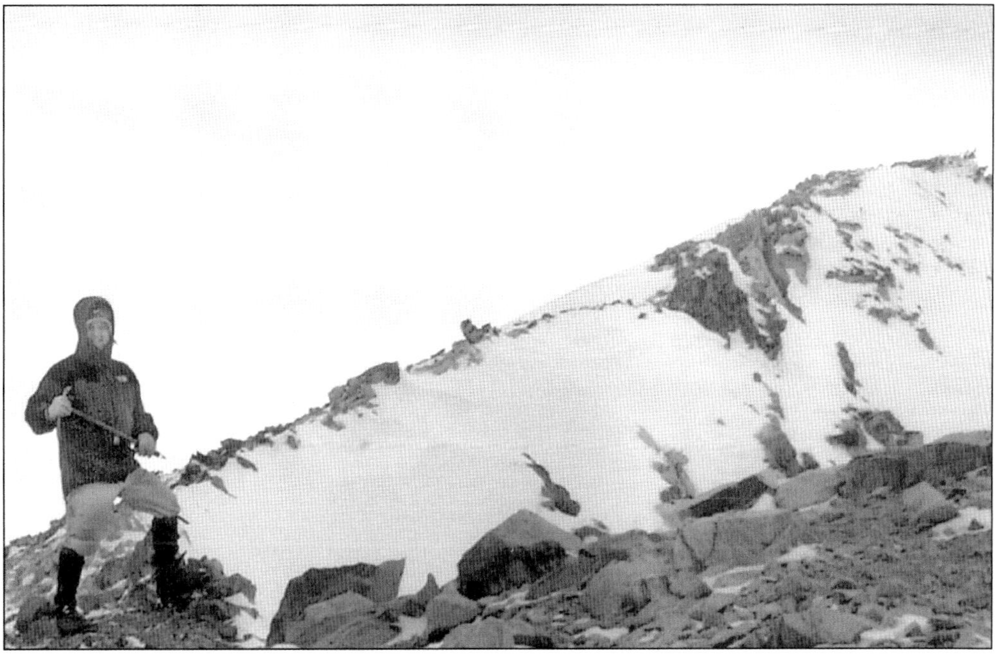

Bob Blanck hikes along the back of the Dragon, the name of the ridgeline south of the Tarn. The Tarn is the highest natural lake in Southern California at around 11,000 feet. In the winter, the Tarn fills with snow, and in the spring, the melting snow forms a lake. Depending on the snow level, the lake may remain full all year, but most years it dries up late in the summer.

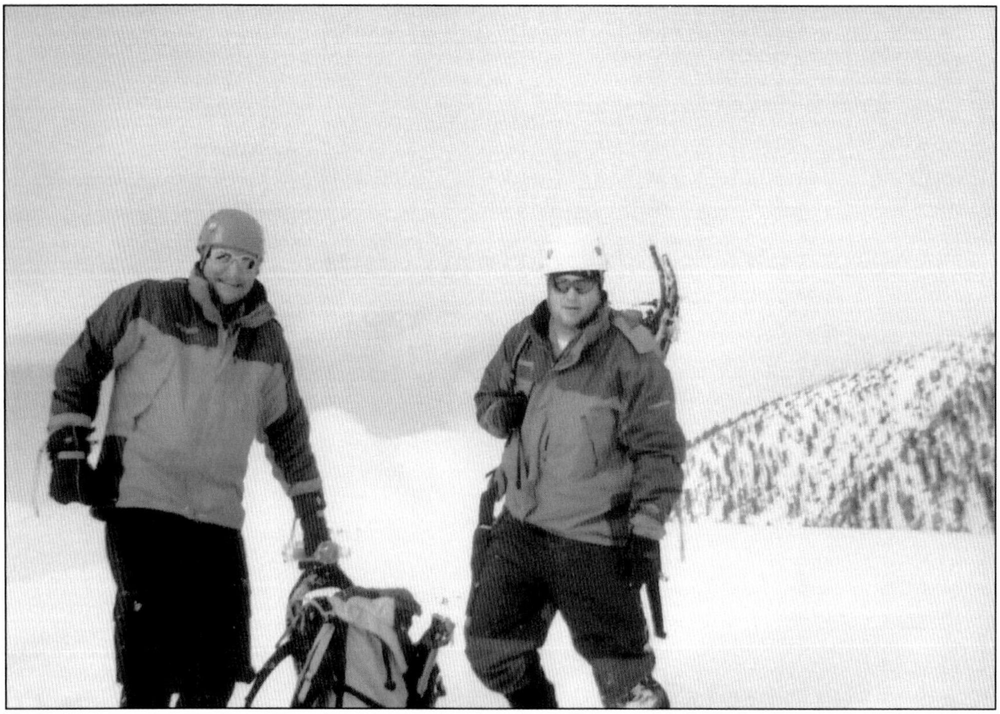

Kevin Newland (above, left), Shane Elliott (above, right), and Bob Blanck (below) are pictured here just after being dropped off by a helicopter near the summit of San Gorgonio in the winter of 2004. The three men were searching for a man from San Diego who had been reported missing the day before. Unfortunately, the missing man was not located alive. He had fallen down an ice chute near Vivian Creek.

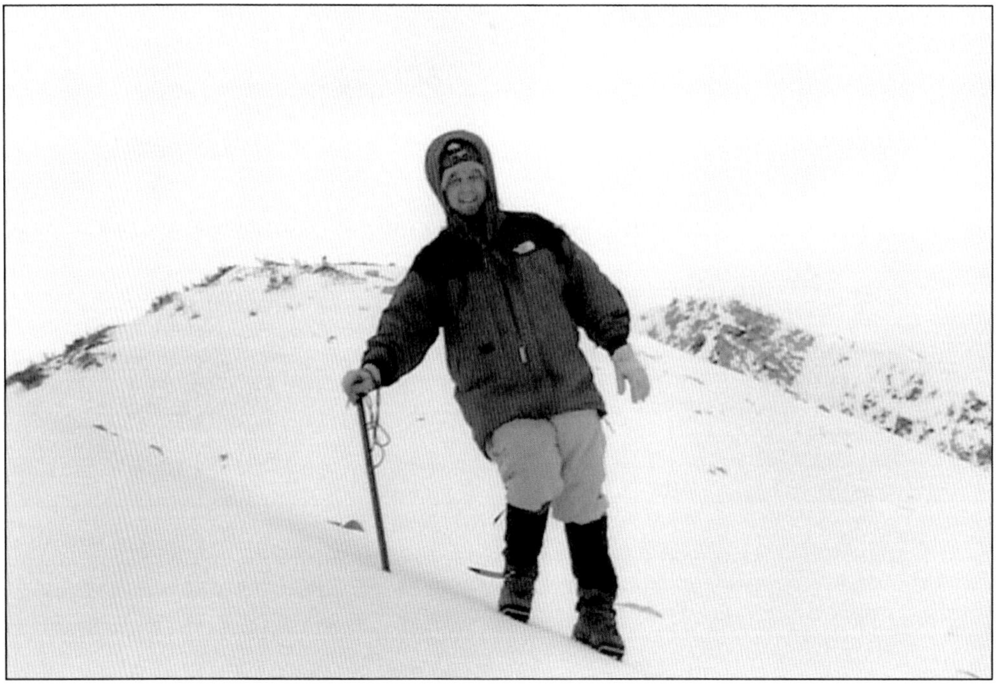

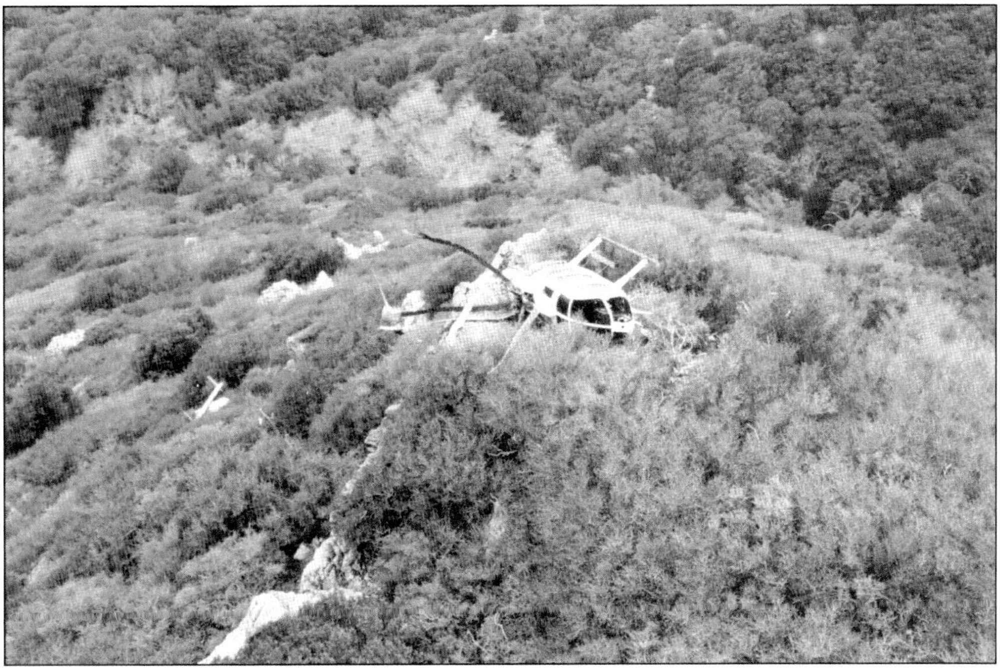

When a sheriff's patrol helicopter was investigating an abandoned vehicle near Monument Peak in the San Bernardino National Forest, a cloud suddenly enveloped it, the pilot lost his bearings, and the helicopter crashed onto a small ledge. The pilot and crew were miraculously uninjured, but they did not know their exact location.

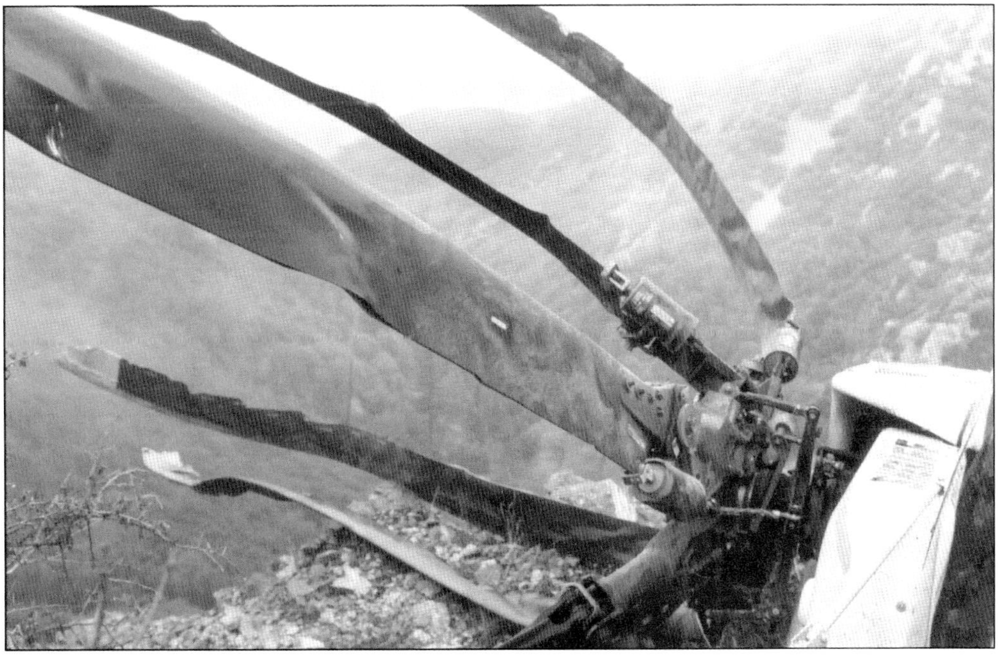

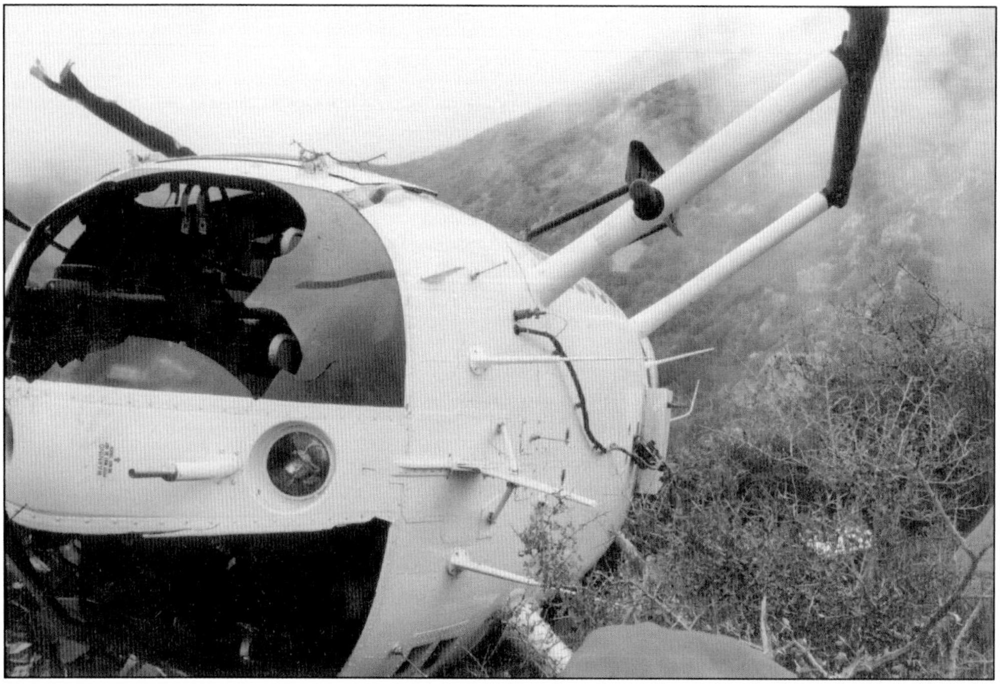

It was San Gorgonio Search and Rescue Team members Rod Walters, Roger Gossett, Scott Johnston, and Bob Blanck who found the helicopter and made their way through the thick manzanita to make contact with the crew. They assessed the crew's medical condition, gave them warm clothes and blankets, and made them comfortable until sunrise.

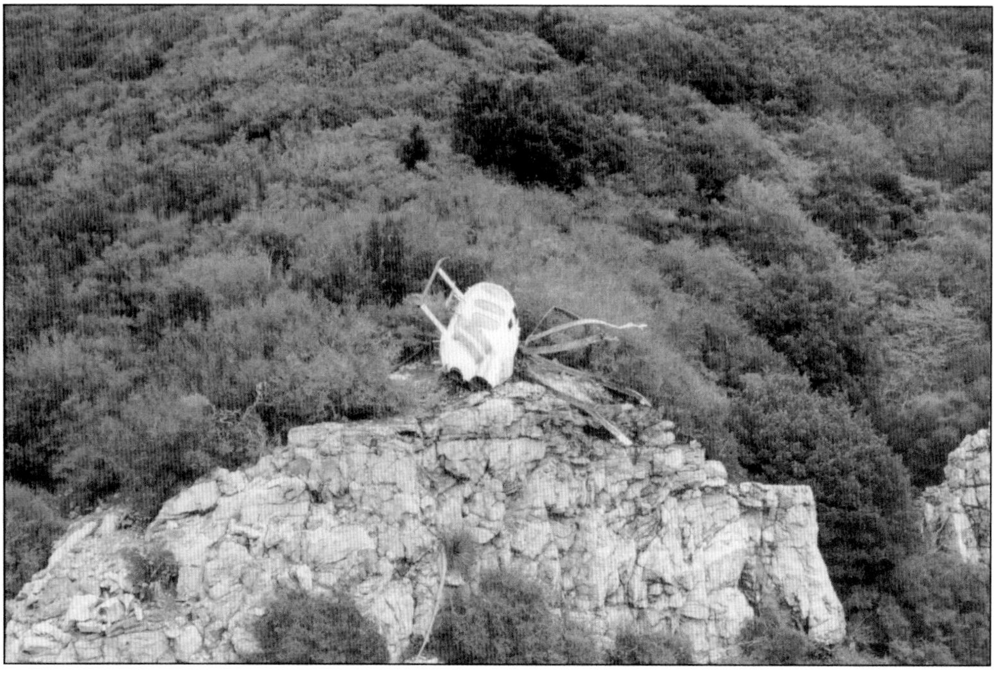

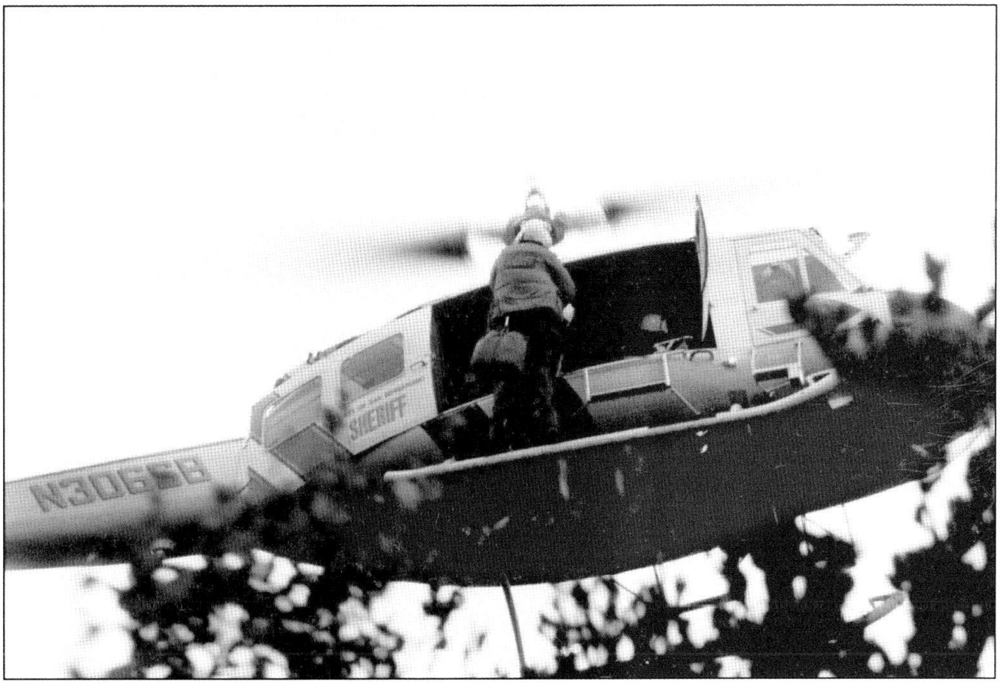

Once the sun rose, the sheriff's department dispatched a rescue helicopter to hoist out the crew and the San Gorgonio team members. Roger Gossett (above), one of the San Gorgonio team members, is happy to be aboard the rescue ship and on his way home after a long night of hiking.

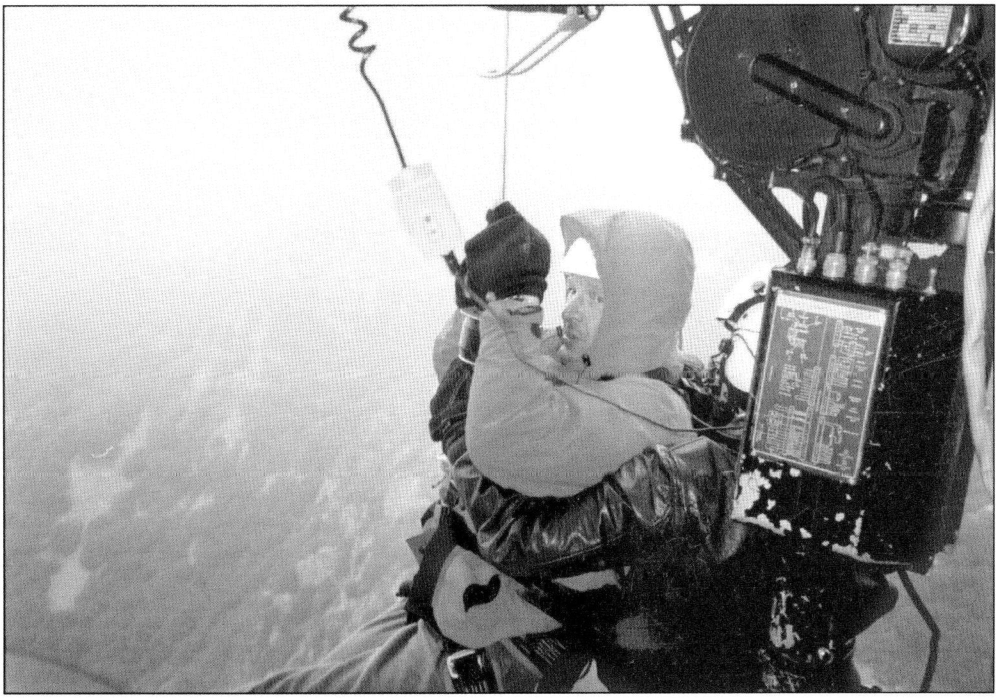

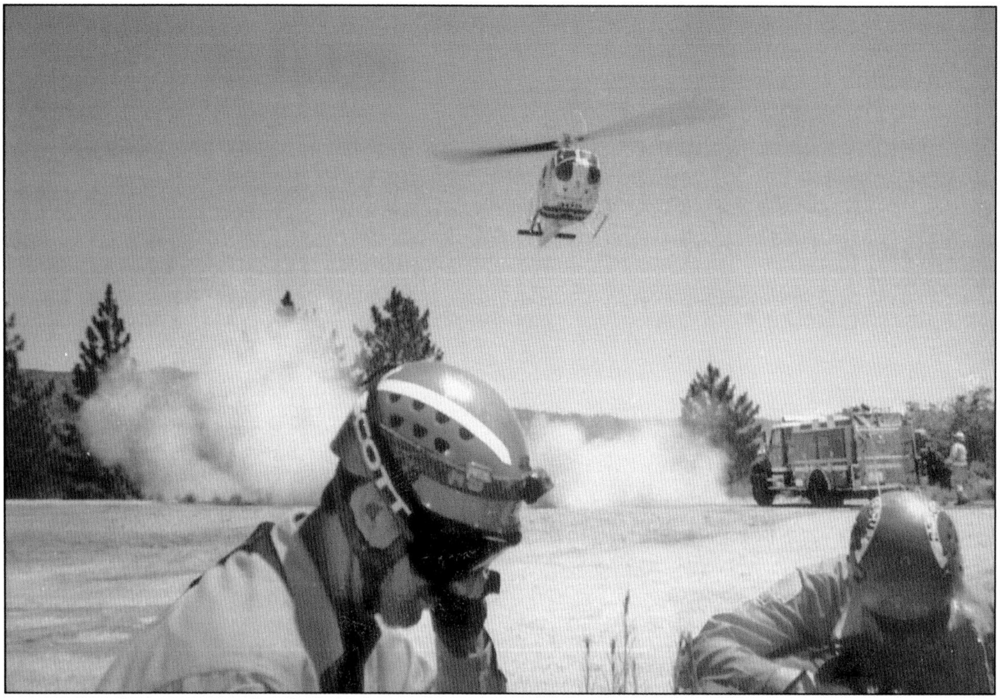

Team members wait as the sheriff's rescue helicopter lands at the landing zone in Angelus Oaks. They are on their way to the summit of San Gorgonio to search for an overdue hiker.

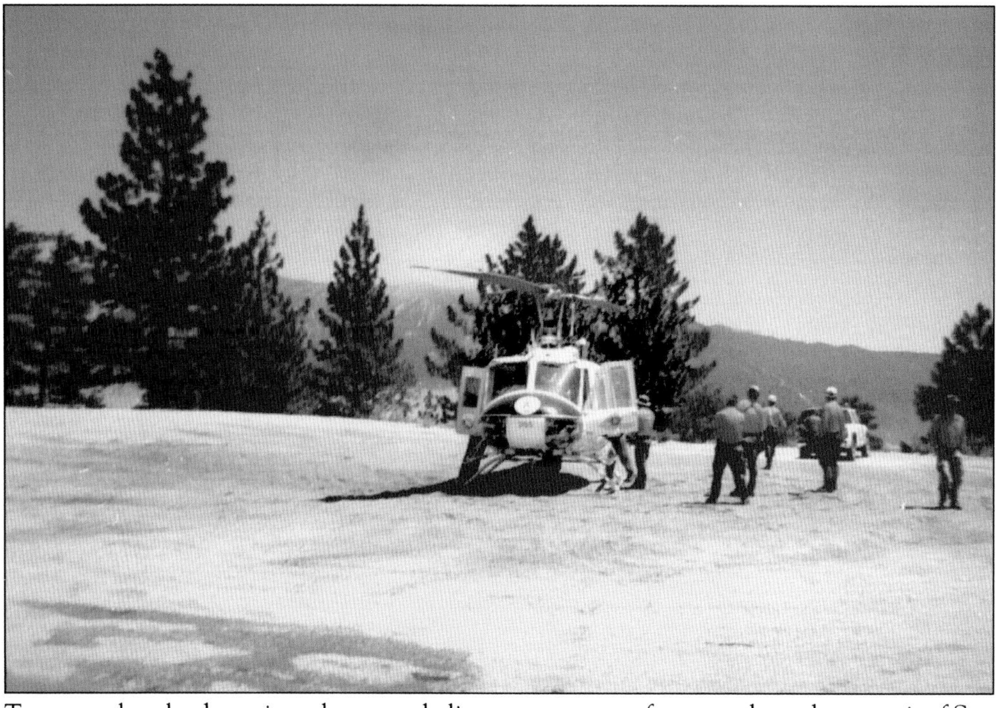

Team members load gear into the rescue helicopter to prepare for a search on the summit of San Gorgonio. When searching for a lost hiker, rescuers are dropped off on the summit and hike downhill and out to the various trailheads in the wilderness.

Six

UNFORGETTABLE SEARCHES AND RESCUES

Over the past 50 years, there are searches and rescues that stand out in the minds of seasoned team members. The San Gorgonio Wilderness has seen its share of celebrities over the years, unfortunately not always with the best outcomes. Frank Sinatra's mother was flying in Sinatra's private jet to Las Vegas one weekend to see her son perform when the plane was caught in a fierce snowstorm over the San Gorgonio Wilderness and crashed into the 10,000-foot ridge southeast of Fish Creek Saddle. There were no survivors of the crash, and it was the San Gorgonio Search and Rescue Team that was first on the scene when the storm finally cleared. They recovered the remains of Sinatra's mother and the crew of the jet and received a personal thanks from Sinatra for their effort and care in this unfortunate event.

Ironically, a few years later, the "Rat Pack" would meet the San Gorgonio Wilderness again. Dean Martin's son, Dean Martin Jr., was flying his Air National Guard jet on a training mission over the wilderness when his jet malfunctioned and he crashed. Again, the San Gorgonio Search and Rescue Team was called to recover the body. At the request of the Sinatra and Martin families, no pictures were taken of these recoveries.

Two more high-profile searches that the team took part in are the Laura Bradbury search in Joshua Tree National Park and the Jared Negrete search in the San Gorgonio Wilderness. More recently, the team also played a significant role in the evacuation of residents during the "Old" Fire in 2003, one of the biggest fires to hit the San Bernardino National Forest in decades.

Finally, seeing team members injured or killed in the line of duty is always the most tragic experience any of the group can face. In January 1998, Tom Rutledge fell almost a 1,000 feet down an ice chute on Mount San Jacinto after recovering the body of hiker, and in July 2004, while en route to a search in Joshua Tree National Park, a drunk driver hit one of the search-and-rescue team vehicles head on, killed two of the team members, and injured several more.

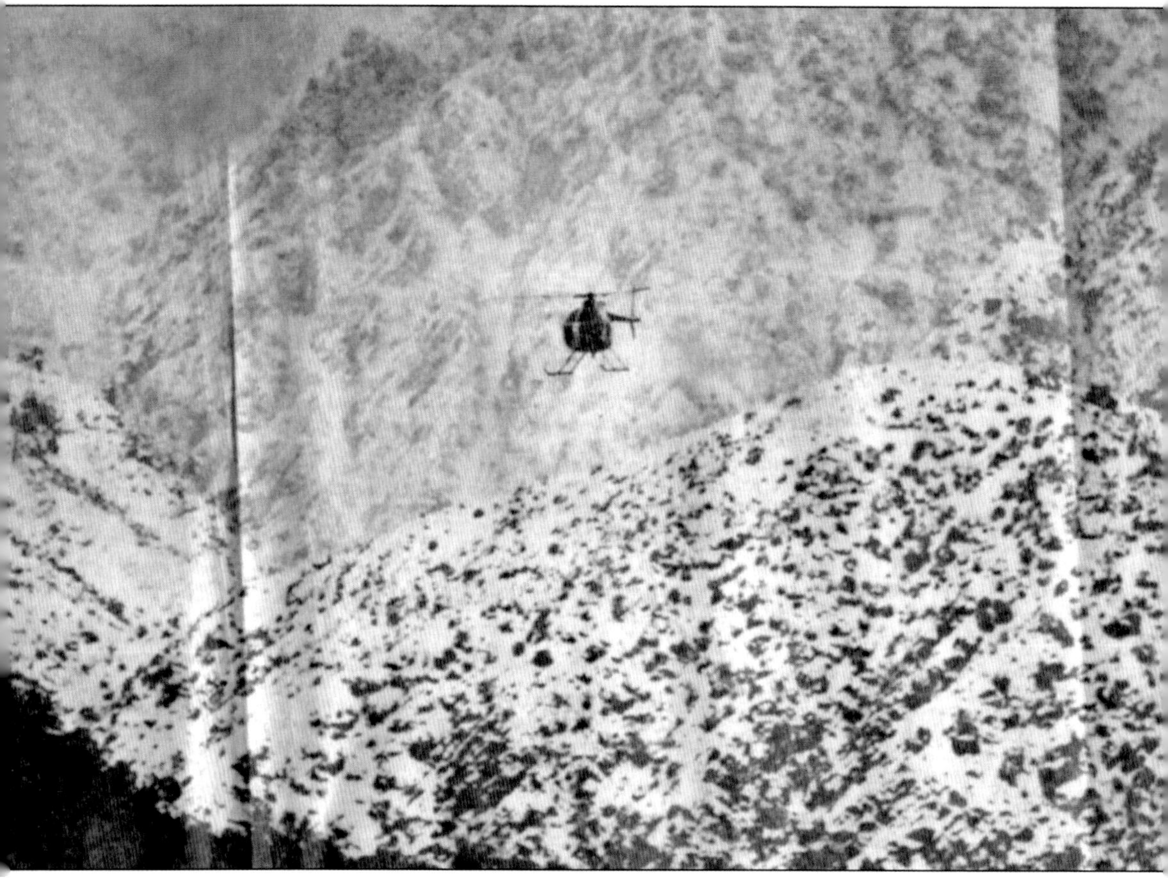

In January 1977, a Lear Jet carrying Frank Sinatra's mother, Natalie "Dolly" Sinatra, crashed into 10,000 Foot Ridge in the remote eastern area of the San Gorgonio Wilderness. The jet had just taken off from Palm Springs Airport and was heading to Las Vegas so Dolly could see her son perform. The search efforts were hampered by extremely bad winter weather, making the white jet difficult to see in the snow. Team members remember seeing Frank Sinatra flying over the crash site in a chartered helicopter. (Article courtesy of *Redlands Daily Facts*.)

Dean Martin's son missing

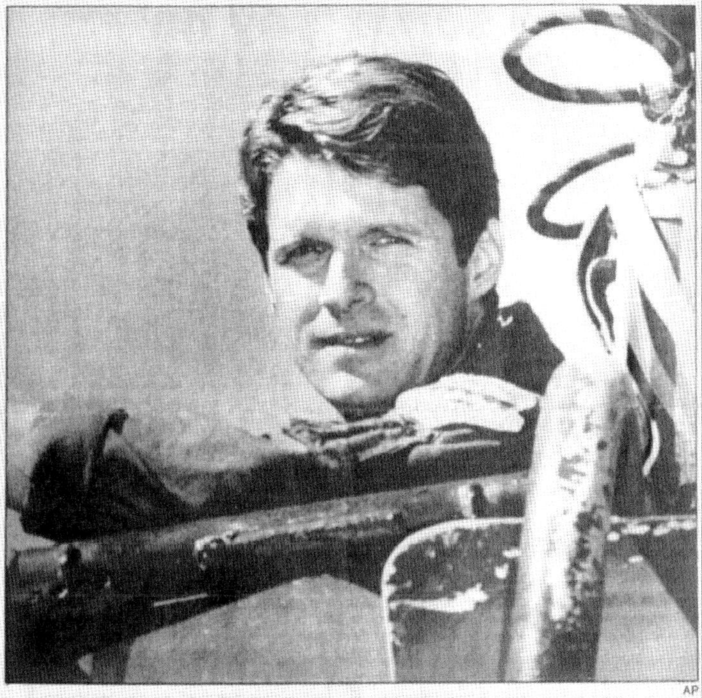

AP

LOVE OF FLYING: Dean Paul Martin, photographed last year in the cockpit of an F-4C Phantom jet, liked 'the thrill' of flying and 'serving his country,' says fellow actor Stephen Lee.

Air Guard fighter crashes in S. Calif.

'Needle in a haystack' search is on

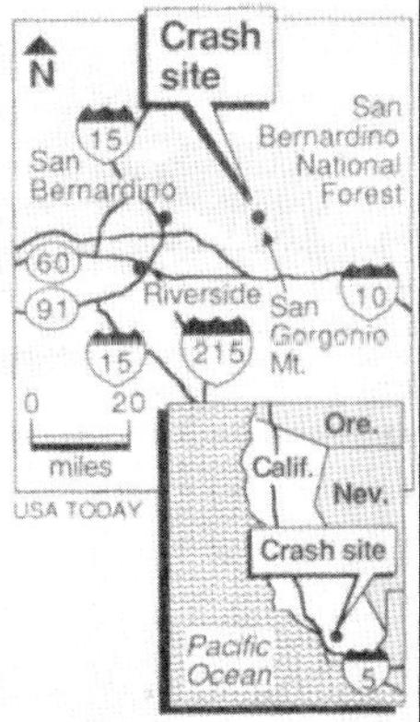

By Susan Manuel
USA TODAY

Actor and athlete Dean Paul Martin — son of entertainer Dean Martin — was still missing late Monday after the crash of his Air National Guard fighter jet on Southern California's highest peak Saturday.

Planes searched the snow-covered 11,500-foot Mt. San Gorgonio for Martin, a 35-year-old captain and pilot and Capt. Ramon Ortiz, 39, a Las Vegas bank employee.

"We are looking for a needle in a haystack," said National Guard Maj. Steve Mensik.

However, rescuers did find the wreckage of another small plane with two survivors.

Actor Stephen Lee, who worked with Martin on a recent TV pilot show, called Martin "one of the sweetest people I've ever worked with."

Martin downplayed his Beverly Hills upbringing and had many passions outside acting.

One major love: flying.

"He just liked the thrill," said Lee. "And I think he was into serving his country."

"Capt. Martin was . . . an exceptional athlete who handles himself exceptionally well in the cockpit," said Mensik.

Martin's son, Alexander, 14, watched the F-4C Phantom jet take off from March Air Force Base, 70 miles east of Los Angeles. Ten minutes later, the plane disappeared from radar.

Martin is divorced from ice skater Dorothy Hamill and actress Olivia Hussey, Alexander's mother.

He played in the professional tennis circuit and tried out with the Los Angeles Rams.

At 14, he formed the rock music band Dino, Desi and Billy with Desi Arnaz Jr. and Billy Hinsche.

In March 1987, a McDonnell Douglas F-4C piloted by Capt. Dean Paul Martin, son of actor and entertainer Dean Martin, crashed into the San Gorgonio Wilderness. The search for the plane was hampered by foul weather with especially strong winds and snow. The wreckage was finally spotted by a sheriff's helicopter several days later. This too is a very memorable mission to the active members at the time because of its celebrity nature and the massive efforts that were put into finding the plane. (Article courtesy of *Redlands Daily Facts*.)

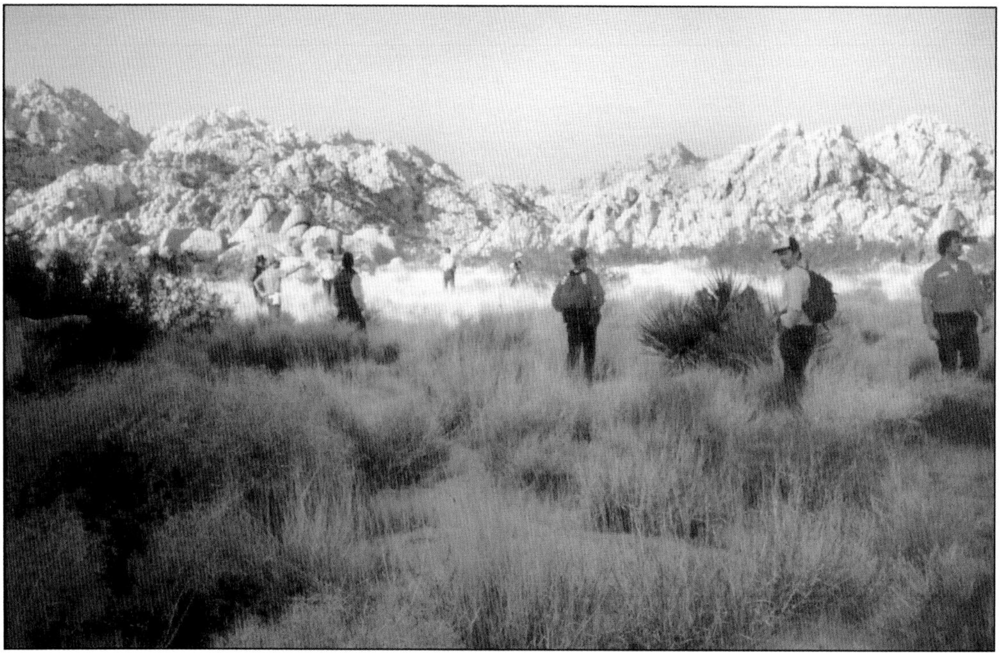

In 1984, four-year-old Laura Bradbury disappeared from her family's campsite in Joshua Tree National Park, and rescue teams from all over the county were called to help search for her. The San Gorgonio team had to adjust some of their search tactics; searching in the mountains was different than searching in the desert environment of Joshua Tree.

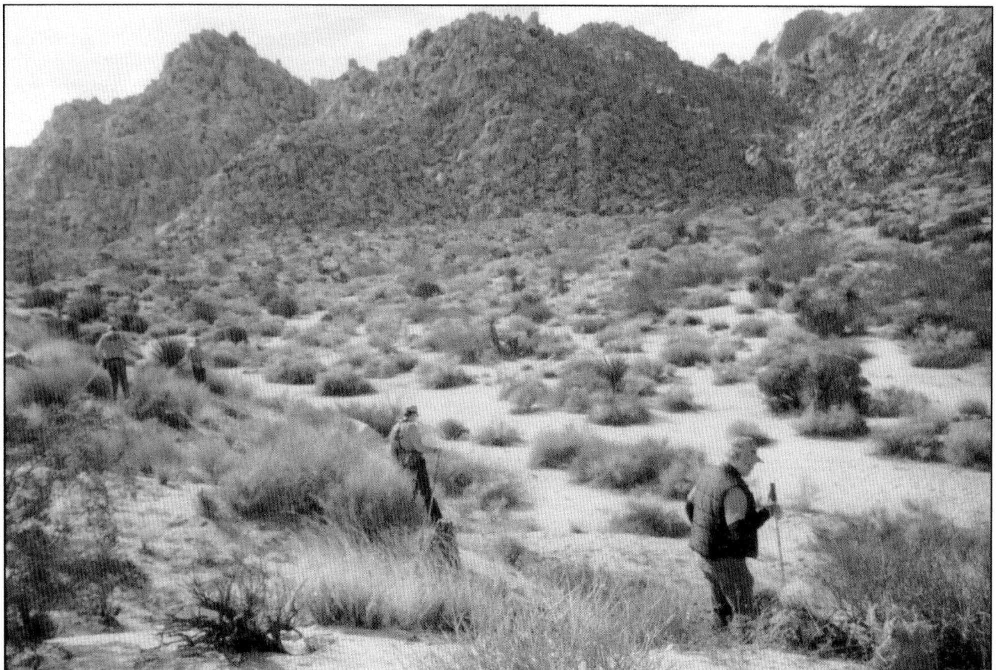

Laura's death may always be a mystery

By MARTHA GUERRERO
Sun Staff Writer

A chapter was closed Friday in the case of a missing Huntington Beach toddler when detectives announced that a skull fragment found in the San Bernardino desert belongs to 3-year-old Laura Bradbury, who vanished in 1984.

The case officially is a death investigation now, but sheriff's investigators are unsure whether they will be able to answer many questions about how the blonde, blue-eyed girl disappeared from a desert campground.

"This in a sense finishes it," said sheriff's spokesman Jim Bryant. "But yes, those questions re-

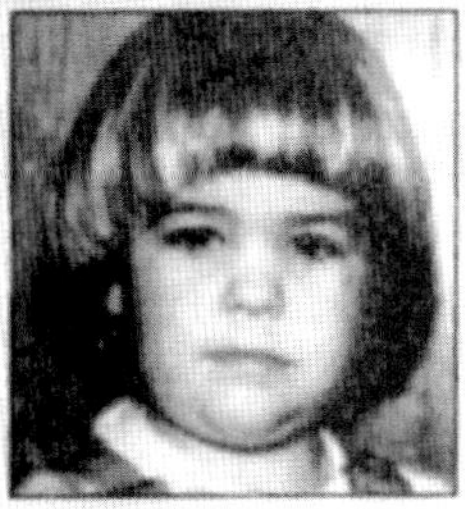

Laura

main. What happened to that little girl and how did she breathe her last breath? Those questions may remain even after the time I'm gone."

Laura Bradbury was 3 years old when she vanished from the Indian Cove Campground at the Joshua Tree National Monument.

The wide-eyed toddler had followed her brother to an outdoor toilet about 50 yards from where the family was setting

See LAURA/Back page

There were many theories about what happened to Laura Bradbury. The family was convinced she had been kidnapped. The search was finally called off after more than of month. Finally, in 1989, a fragment of skull was found that was positively confirmed through DNA testing as that of Bradbury. (Article courtesy of *Redlands Daily Facts*.)

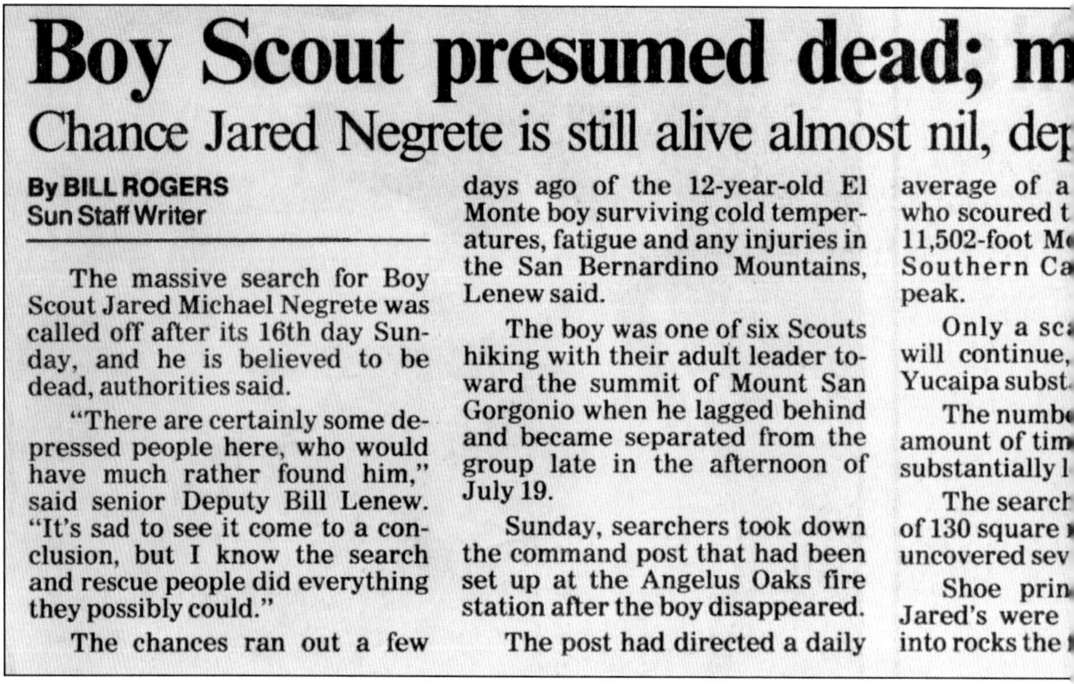

Boy Scout presumed dead; m

Chance Jared Negrete is still alive almost nil, de

By BILL ROGERS
Sun Staff Writer

The massive search for Boy Scout Jared Michael Negrete was called off after its 16th day Sunday, and he is believed to be dead, authorities said.

"There are certainly some depressed people here, who would have much rather found him," said senior Deputy Bill Lenew. "It's sad to see it come to a conclusion, but I know the search and rescue people did everything they possibly could."

The chances ran out a few days ago of the 12-year-old El Monte boy surviving cold temperatures, fatigue and any injuries in the San Bernardino Mountains, Lenew said.

The boy was one of six Scouts hiking with their adult leader toward the summit of Mount San Gorgonio when he lagged behind and became separated from the group late in the afternoon of July 19.

Sunday, searchers took down the command post that had been set up at the Angelus Oaks fire station after the boy disappeared.

The post had directed a daily average of a
who scoured t
11,502-foot M
Southern Ca
peak.

Only a sc
will continue,
Yucaipa subst

The numb
amount of tim
substantially l

The search
of 130 square
uncovered sev

Shoe prin
Jared's were
into rocks the

In the summer of 1990, a Boy Scout named Jared Negrete was separated from his troop while hiking in the San Gorgonio Wilderness. The San Gorgonio Search and Rescue Team was called to begin searching for Negrete. After several days of finding no trace of him, the rest of the county's search-and-rescue teams were called into action. After more than a week of not finding

ssive search ends

y says

200 rescuers
gged slopes of
San Gorgonio,
nia's tallest

down search
the sheriff's
in charge.

eople and the
rching will be

ss large parts
of wilderness,
olid clues.

lieved to be
ed for a mile
eekend.

A second set of prints was followed to no avail three miles away on the fourth day of the search. The last signs of the boy were his camera and beef jerky and dried fruit wrappers, found the eighth day.

Hope had remained high after the film in the boy's camera was developed and showed scenery where the boy may have wandered.

On July 29, Jared's mother, Linda Negrete, joined the search.

She circled above the mountains in a helicopter, calling out to her son over a loudspeaker.

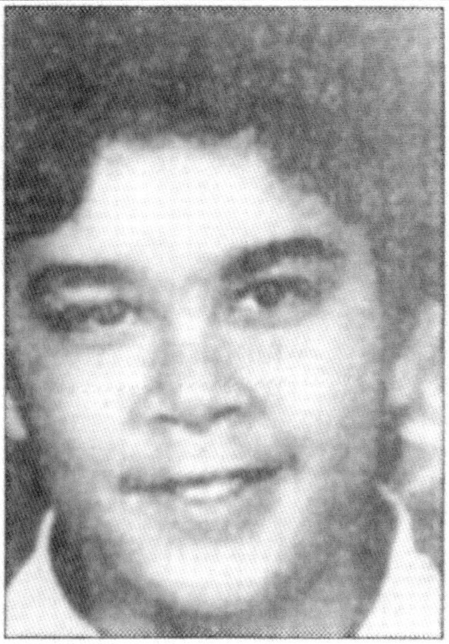

Jared Negrete
Command post disassembled

Negrete, the marines were called in to aid in the search. After more than a month of searching, the search was finally called off. To this day, the Negrete case is the largest unsolved search ever conducted in San Bernardino County. (Article courtesy of *Redlands Daily Facts*.)

Alpine rescuer badly hurt in fall

▶ Angelus Oaks man suffers critical injuries in 1,000-foot slide, after recovery of hiker's body.

By Susan Thurston
The Press-Enterprise

IDYLLWILD

A mountain rescue worker was seriously injured Monday when he slipped and tumbled about 1,000 feet minutes after retrieving the body of a hiker who died after falling off an icy trail the day before, authorities said.

Tom Rutledge, 50, of Angelus Oaks, suffered major head and back injuries after plummeting down an ice chute, San Bernardino sheriff's spokesman Chip Patterson said. He remained unconscious Monday afternoon.

Iskander Toubia, 46, of Irvine, was hiking from Fern Valley near Idyllwild on Sunday morning en route to the lookout tower on top of the Devil's Slide trail.

He was reported missing about 5 p.m. after he could not be reached on his cellular phone, said Riverside County sheriff's Lt. Darryl Birney.

Toubia's body was found at 2 a.m. Monday about 300 feet down from the trail just north of the lookout tower by rescuers with the Riverside Mountain Rescue Unit and the San Gorgonio Search and Rescue Team.

At daylight, a sheriff's helicopter was brought in to retrieve the body of Toubia, who lived in Fern Valley on weekends.

The body was flown to the

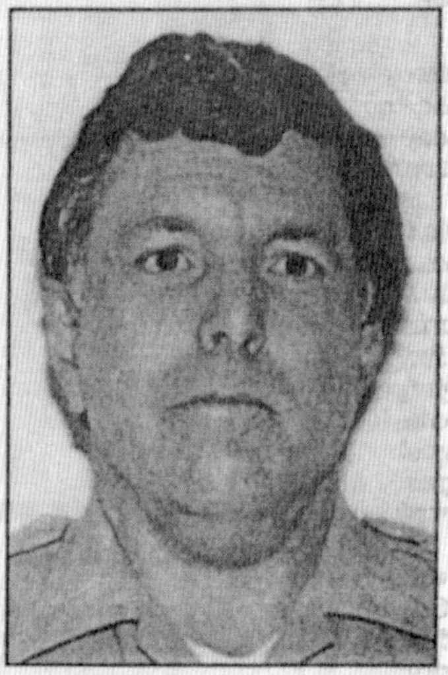

Ton Rutledge

Keenwild Air Base in Mountain Center at 11:15 a.m. Five minutes later, Rutledge fell while preparing to leave.

Rutledge was taken to Desert Hospital in Palm Springs, where he was in critical condition Monday evening.

Rutledge has been a volunteer with the San Gorgonio team for about 18 months, Patterson said. The team is part of the San Bernardino County Sheriff's Department.

Packy Wassell, a 30-year member of the San Gorgonio team, said Rutledge was a volunteer firefighter in Angelus Oaks and owns the Oaks Restaurant in Angelus Oaks.

"It's very steep country — a lot of rock and steep terrain. You've got snow on the ground and footing is extremely dangerous," he said. "It's extremely tough to get around and you have to be careful."

Unfortunately, the team faces tragedy with its own members as well. After successfully completing the body recovery of a hiker who had fallen down an ice chute on Mount San Jacinto, team member Tom Rutledge accidentally slipped and fell down the same ice chute. Tom slid head first on his back more than 1,000 feet down the chute. He suffered massive head injuries and went through over a year of rehabilitation. (Article courtesy of *Redlands Daily Facts*.)

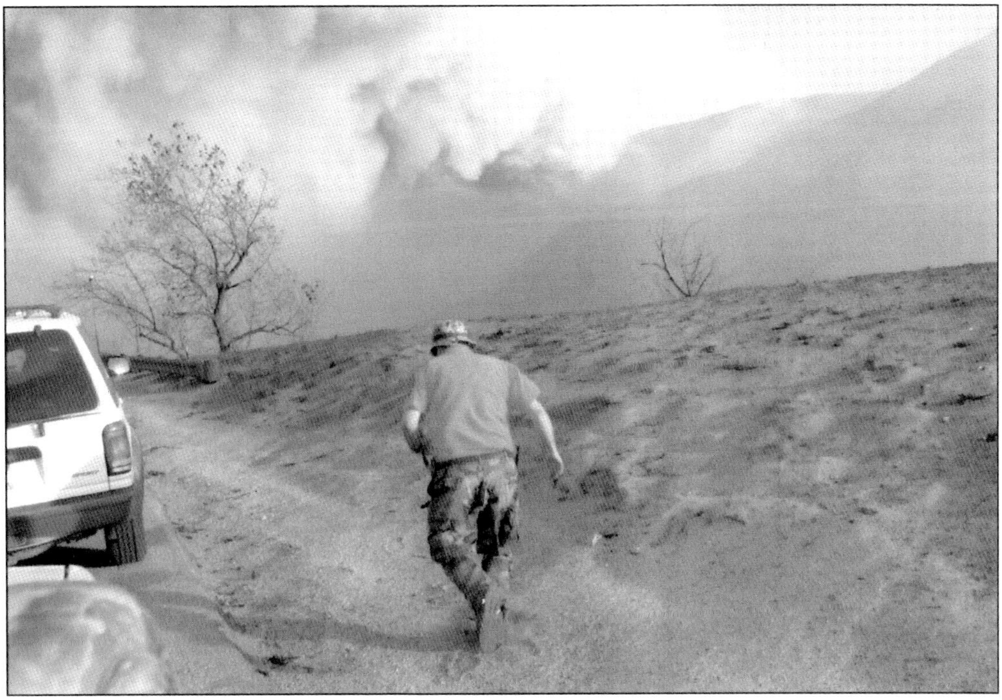

Aside from rescuing lost hikers, in recent years, the team has been asked to assist the county fire departments with evacuations of local neighborhoods threatened by fire. The team will go to each house, inform residents of the eminent danger, and help them evacuate. Then team members will stay behind and man road blocks to keep residents from returning to their homes until the danger is clear.

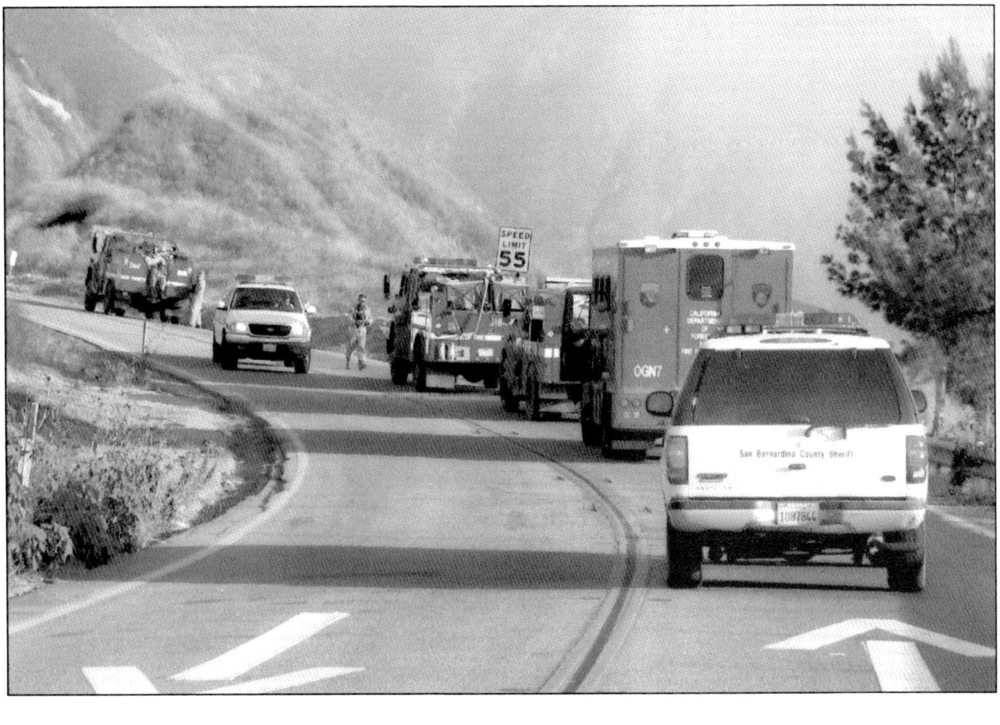

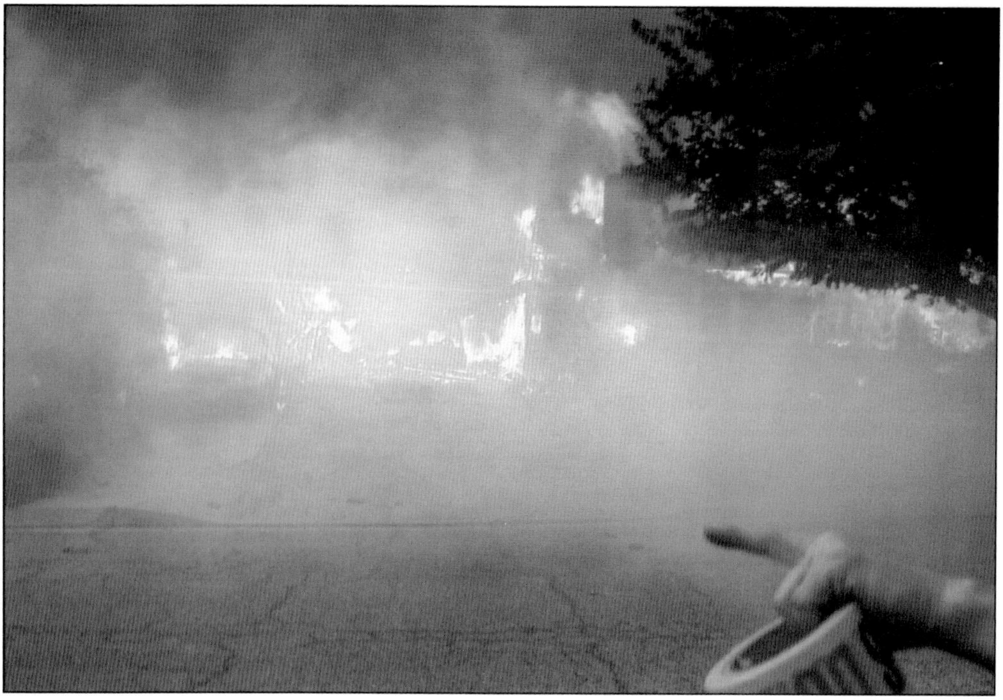

In October 2003, the "Old" Fire (so named because it was started by an arsonist on Old Waterman Canyon Road) swept up Highway 18 and Waterman Canyon destroying houses along the way.

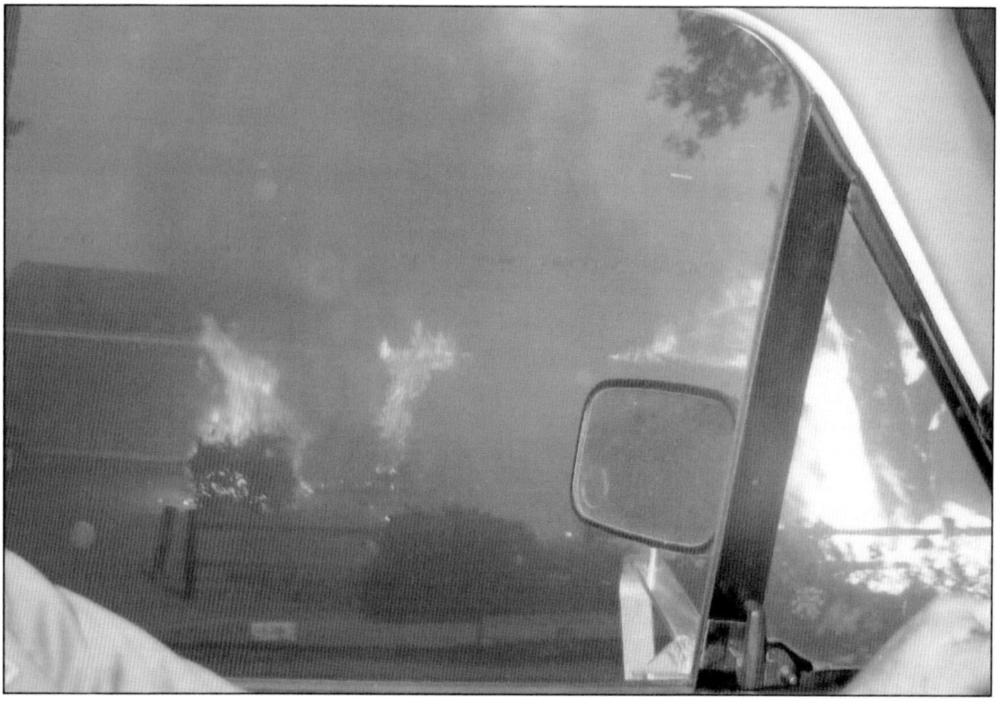

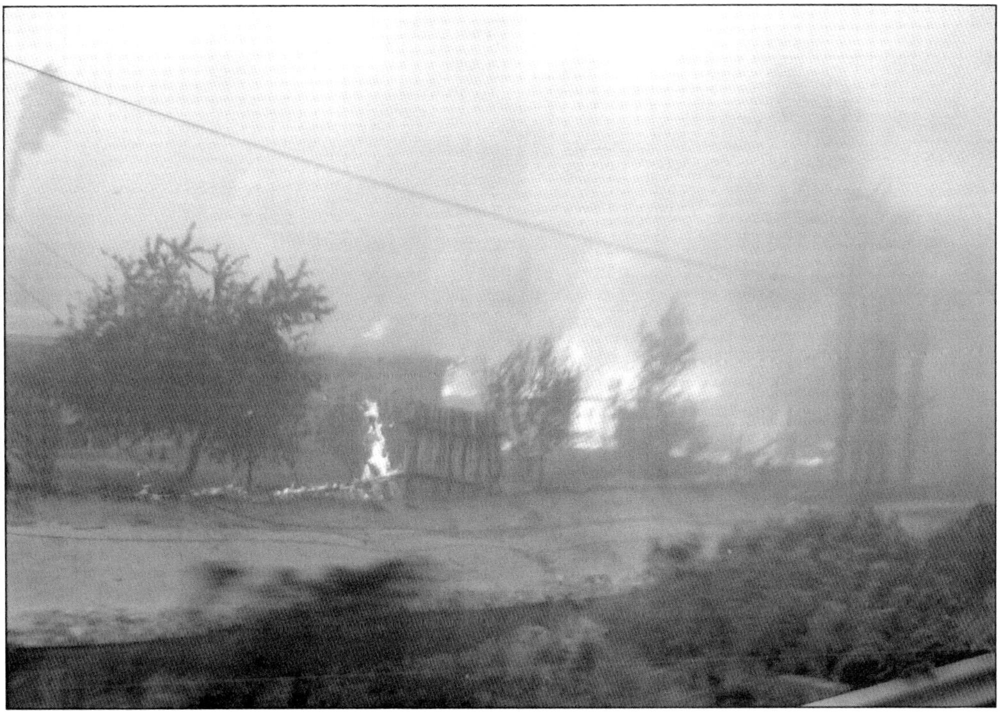

The fire eventually made it into parts of Crestline, Rim Forest, and Running Springs before the winds changed and blew the fire back down the mountain into parts of Highland and San Bernardino.

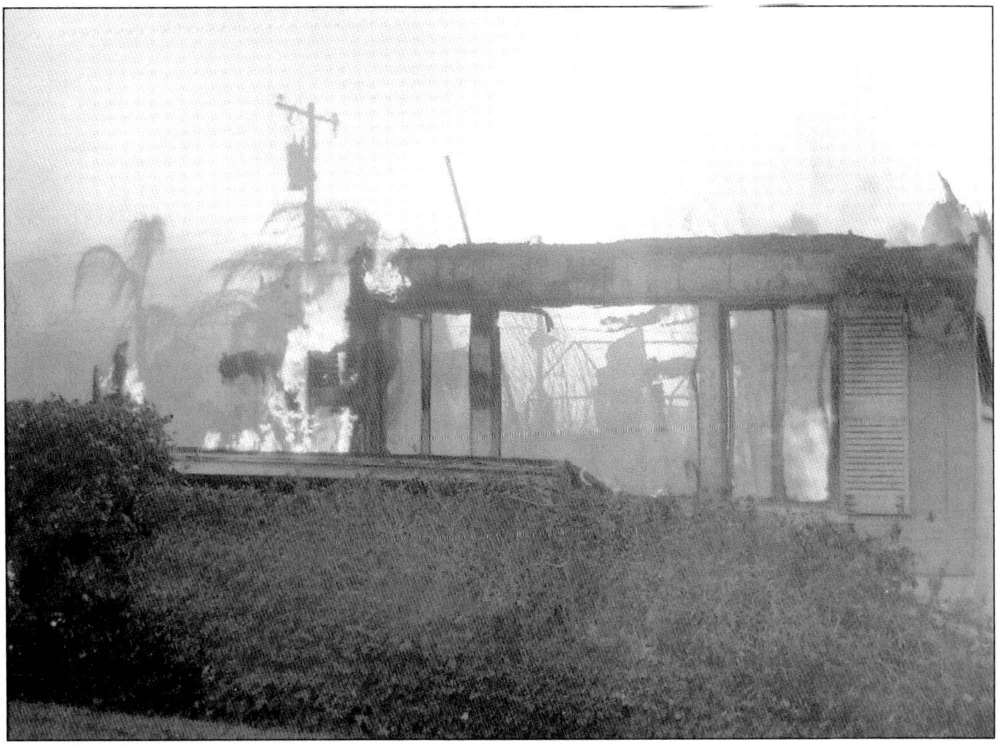

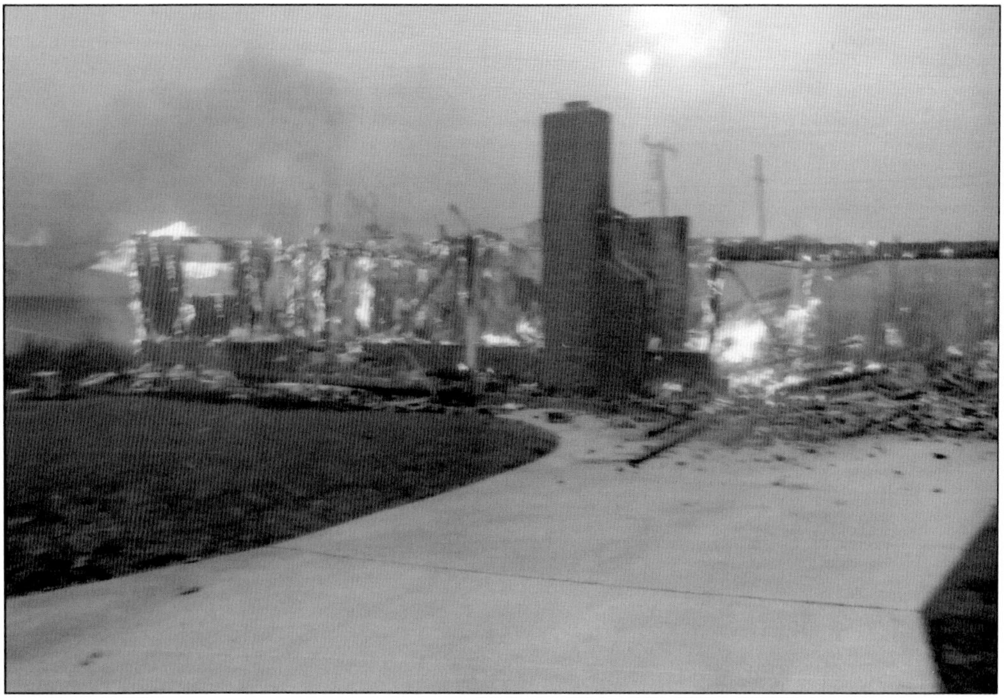

Even though this is not the San Gorgonio Search and Rescue Team's area, the San Bernardino County Sheriff's Department activated all of its volunteer units to aid in the evacuation of residents in areas threatened by the fire.

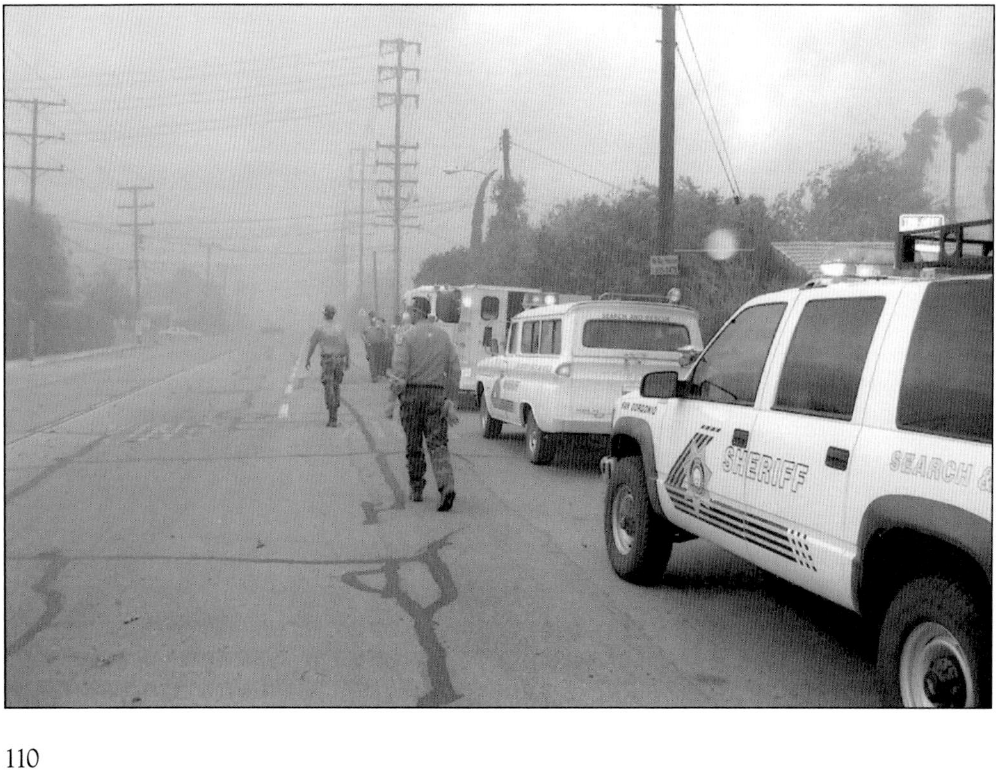

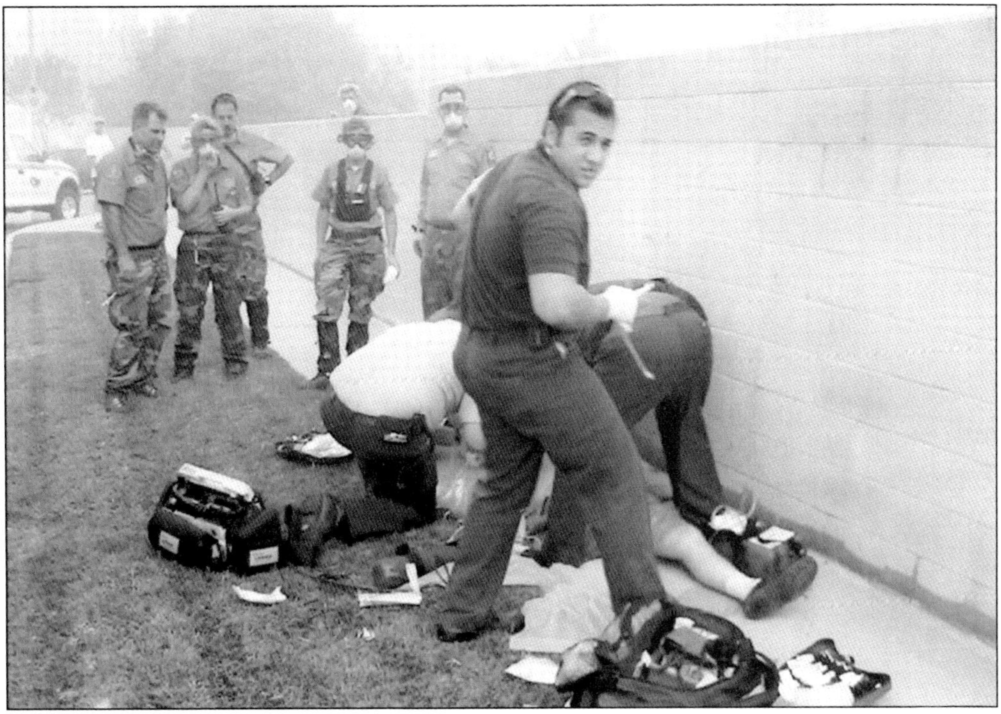

An elderly man was frantically watering down his burning roof, but the search-and-rescue crew convinced him it was not safe to stay there and brought him back to the command area. Shortly after that, he went into cardiac arrest from the stress of losing his house. Team members, and later paramedics, performed CPR to revive him, but unfortunately, he died at the command area.

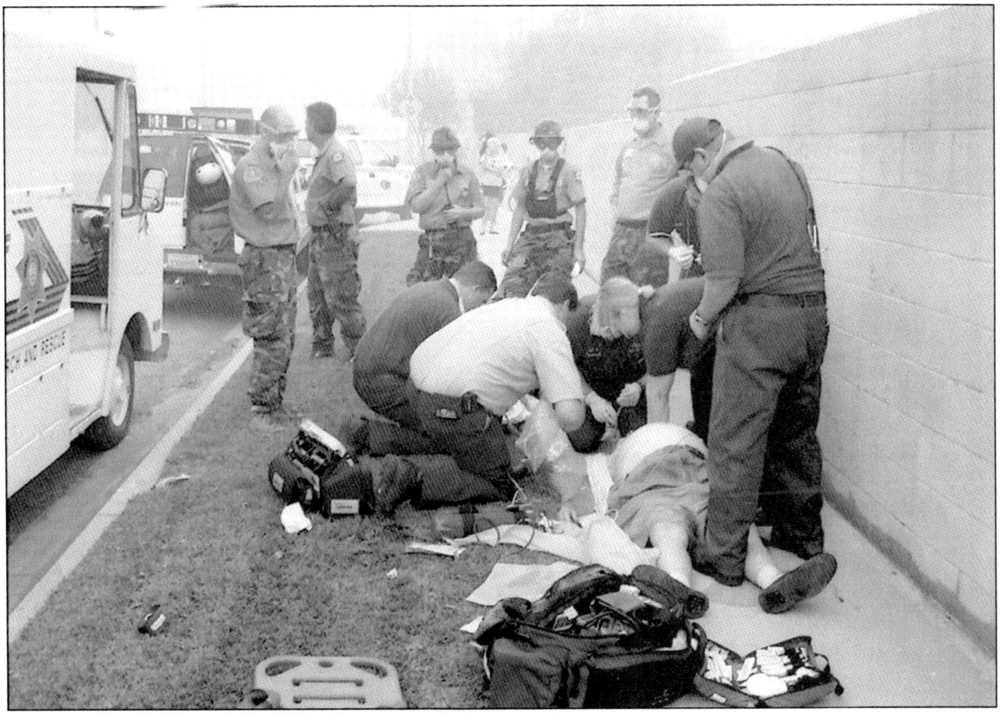

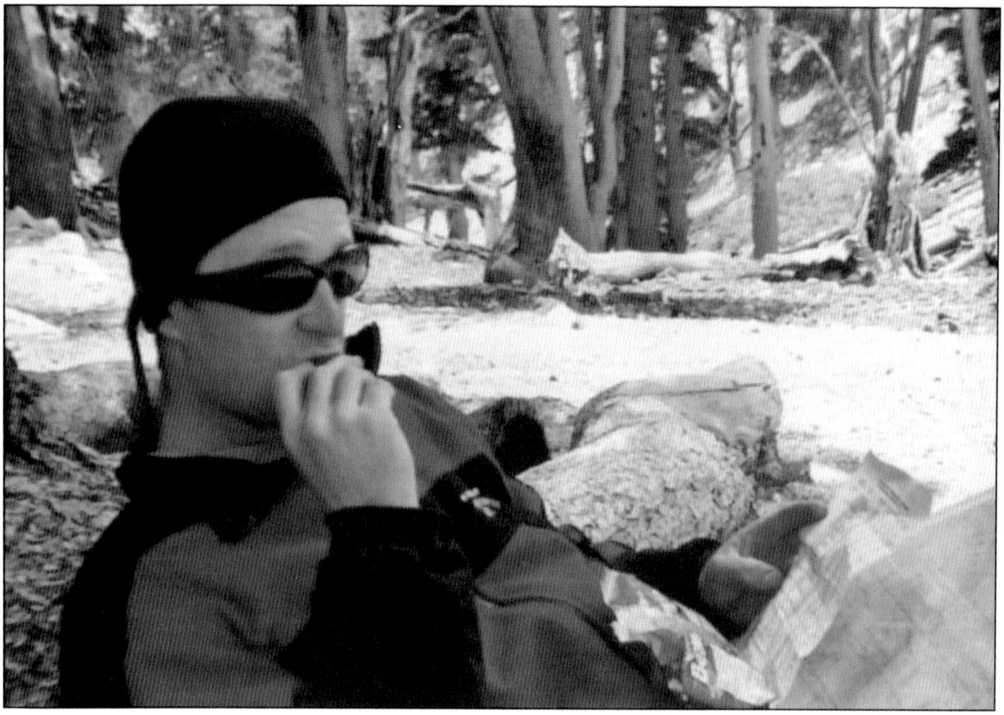

On July 17, 2004, the team was called to Joshua Tree National park to search for a lost hiker. The team left Yucaipa at 4:00 a.m. At 5:00 a.m., while driving down Highway 62 through Morongo Valley, California, the team Suburban was struck head on by a drunk driver, killing team member Scott Johnston, pictured above and below.

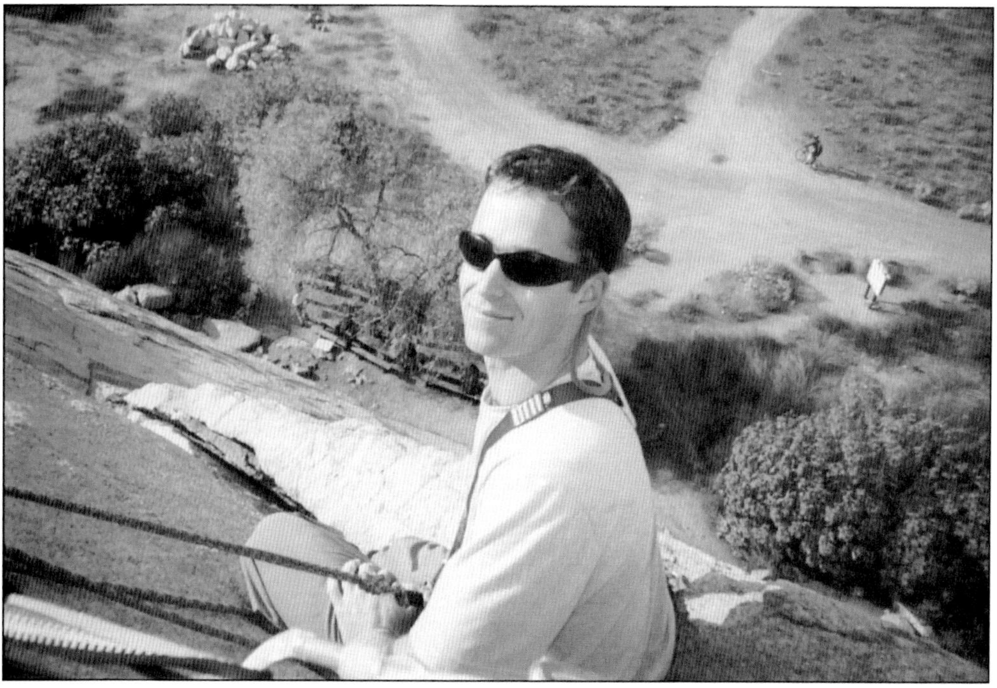

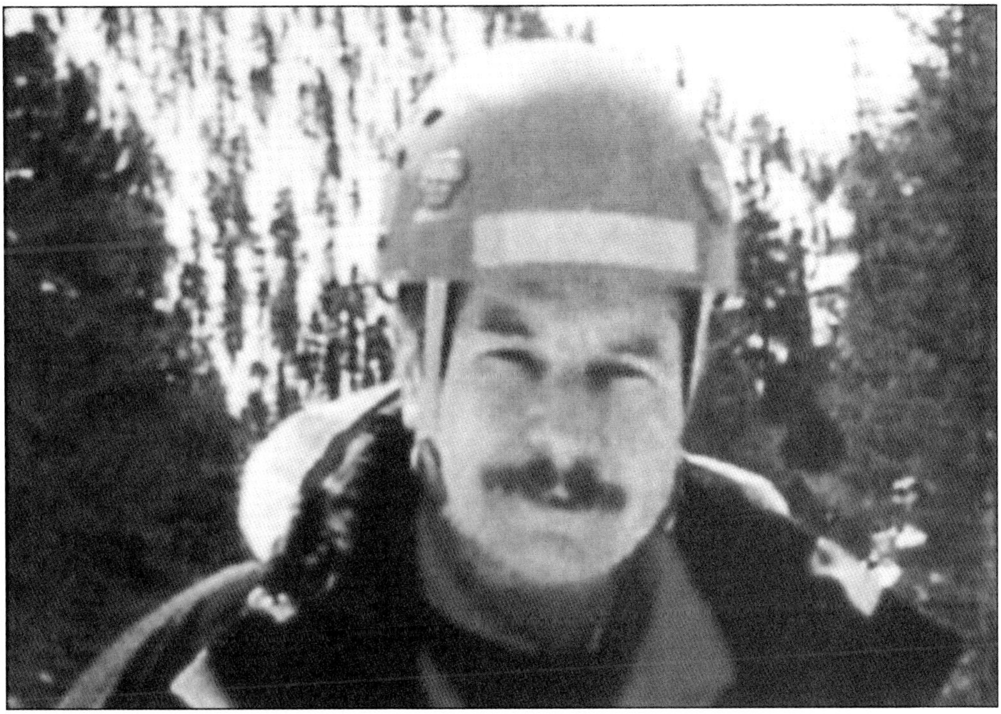

Team member Phil Calvert, pictured above and below, was seriously injured and rushed to Desert Regional Hospital. He was treated for massive internal injuries and broken bones, but he died of his injuries two days after the accident.

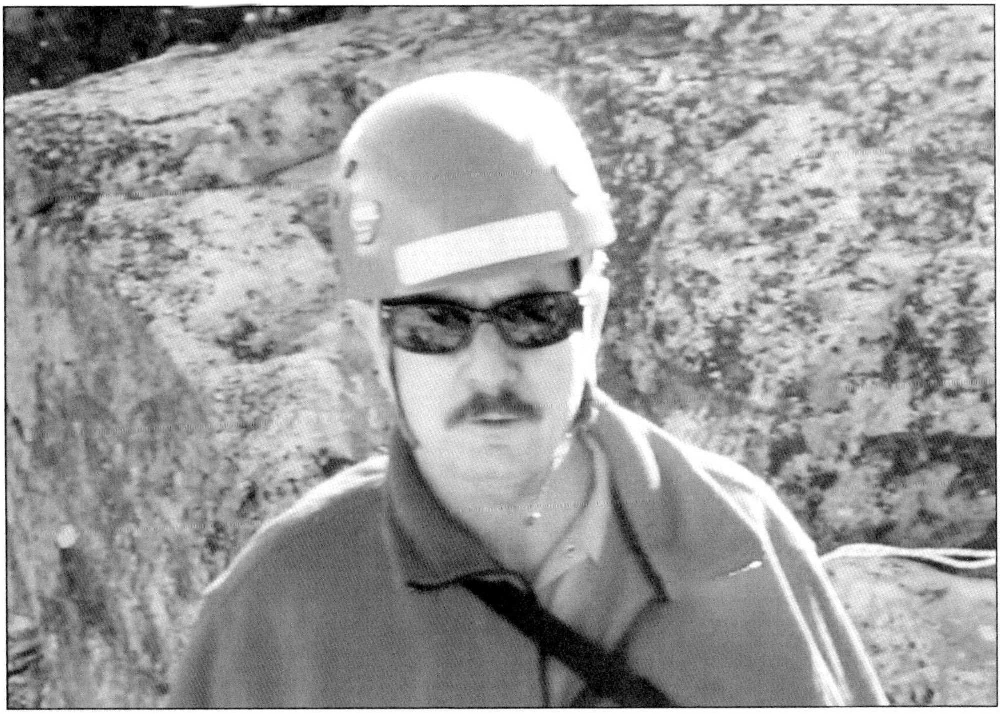

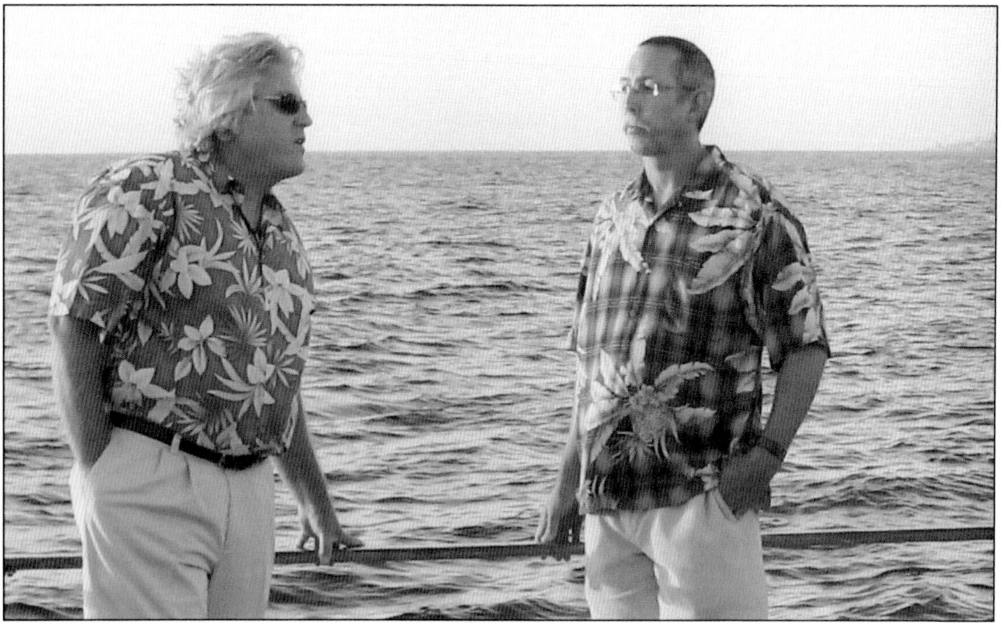

Team members were faced with the reality of saying good-bye to two of their teammates and closest friends within the same week. Team members Walt Farmer and Ron Dietzman (pictured above) tell stories about Scott after he was laid to rest at sea. Pictured below, team members Ellyn Loenhorst and Julie Haggerty along with Julie's daughter Jennifer remember Scott and Phil as they watch the sunset after Scott's memorial at sea.

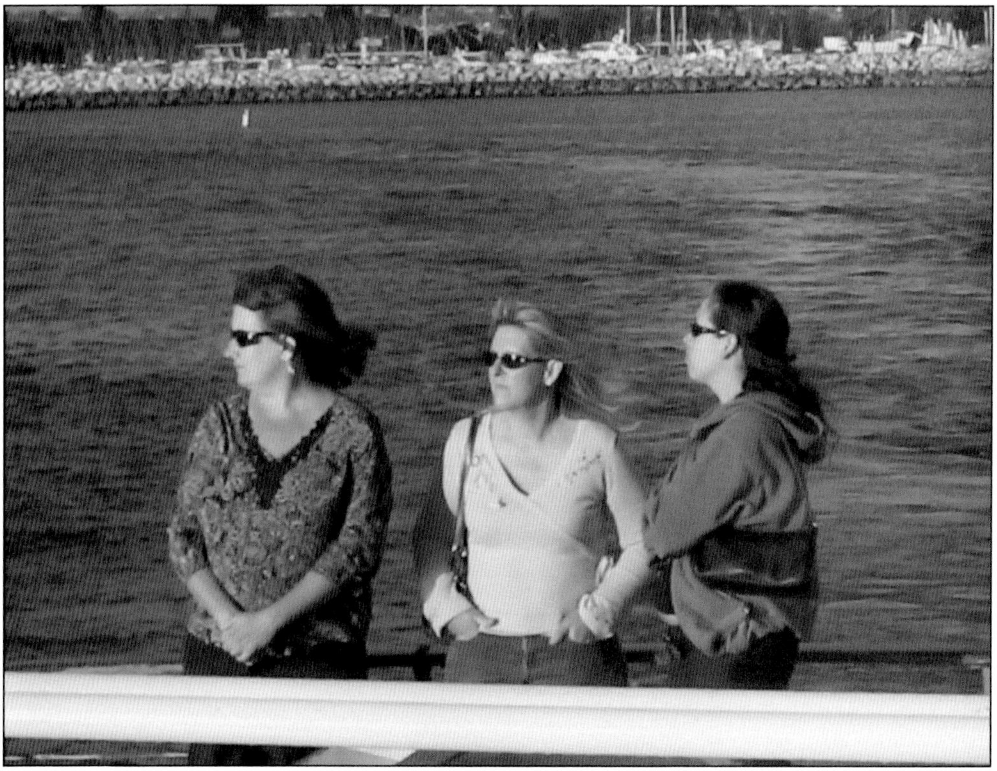

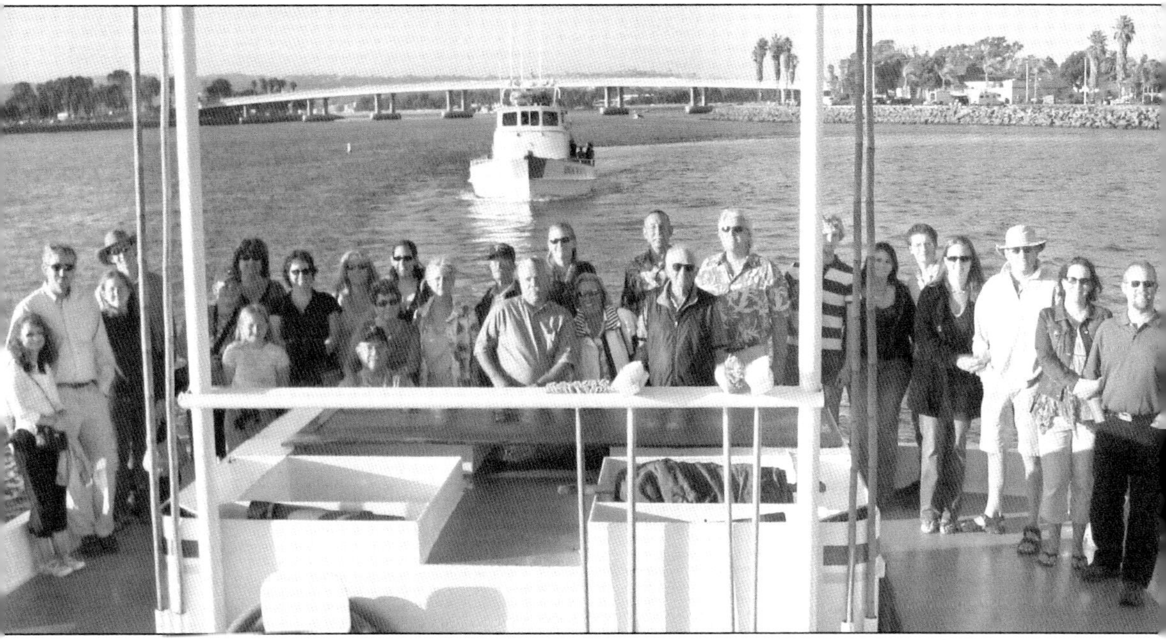

Friends and family pose for a picture on the boat after saying good-bye to Scott Johnston at sea just off the coast of San Diego, where he loved to surf and scuba dive. Besides Scott's memorial at sea, Phil and Scott both had memorial services at local churches in Redlands, California. Both were organized and conducted by the sheriff's office, with all the rites and privileges due a volunteer who was killed during a search and rescue mission.

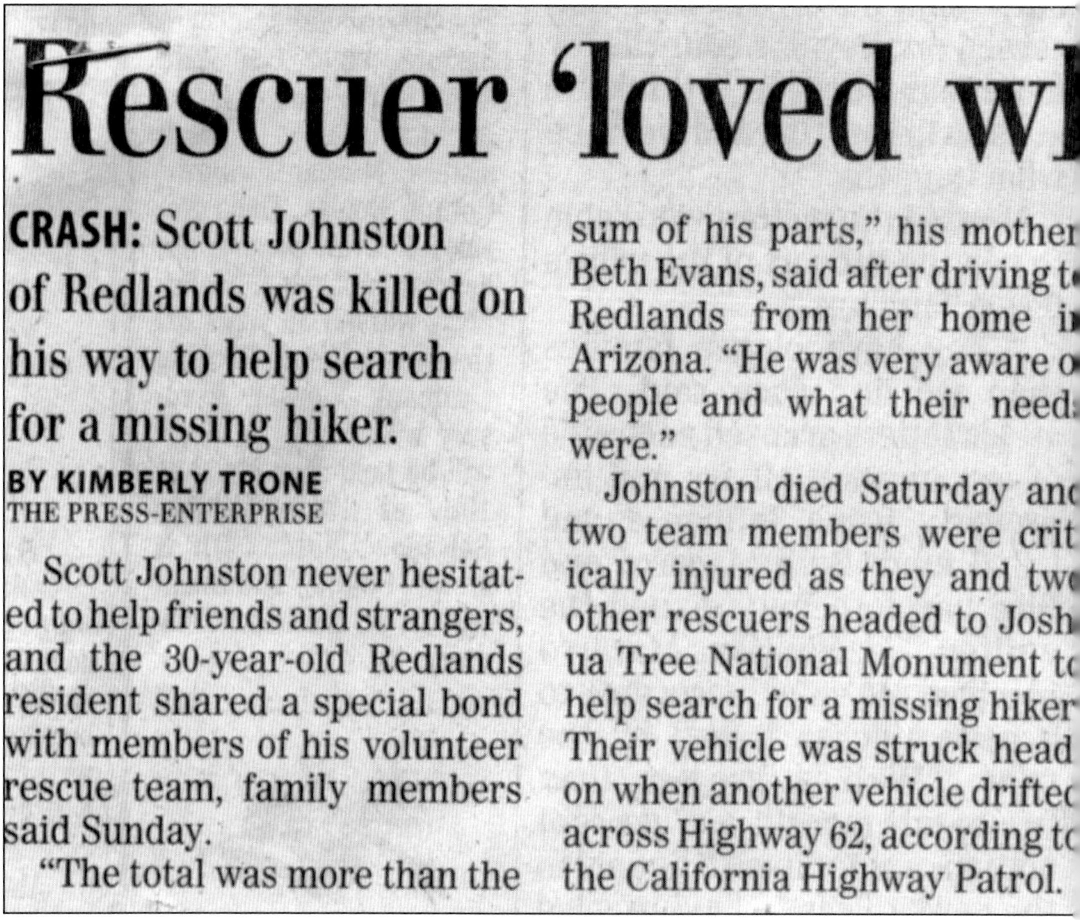

Rescuer 'loved w

CRASH: Scott Johnston of Redlands was killed on his way to help search for a missing hiker.

BY KIMBERLY TRONE
THE PRESS-ENTERPRISE

Scott Johnston never hesitated to help friends and strangers, and the 30-year-old Redlands resident shared a special bond with members of his volunteer rescue team, family members said Sunday.

"The total was more than the sum of his parts," his mother Beth Evans, said after driving t Redlands from her home i Arizona. "He was very aware o people and what their need were."

Johnston died Saturday an two team members were critically injured as they and tw other rescuers headed to Josh ua Tree National Monument t help search for a missing hiker Their vehicle was struck head on when another vehicle drifte across Highway 62, according t the California Highway Patrol.

Scott was killed instantly when a drunk driver plowed into the search-and-rescue Suburban en route to a search in Joshua Tree National Park. Dr. Jim Evans, a member of the Valley of the Falls Search

at he was doing'

Two of Johnston's team members remained hospitalized Sunday in critical condition.

Phil Calvert was at Desert Regional Medical Center in Palm Springs. Calvert was expected to be on a ventilator at least through today with multiple internal injuries and a broken leg, said Cindy Beavers, San Bernardino County sheriff's spokeswoman.

Another rescue volunteer, Bob Blank, was at Arrowhead Regional Medical Center in Col-

PLEASE SEE **JOHNSTON, B3**

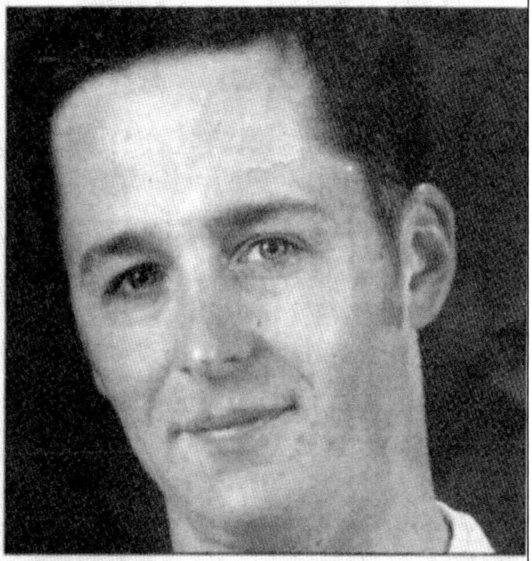

Scott Johnston was a member of a highly specialized team, a sheriff's spokeswoman said.

and Rescue Team and good friend of Scott, attempted to resuscitate him for over 10 minutes and finally pronounced him dead at the scene. (Article courtesy of *Redlands Daily Facts*.)

Second rescue

Mentone man was injured in head-on crash

By JOE NELSON
Staff Writer

A sheriff's search-and-rescue team member died Tuesday from injuries he suffered in a head-on traffic collision while en route to look for a missing hiker in Joshua Tree National Park.

Phillip Calvert, 58, of Mentone — a member of the San Bernardino County sheriff's San Gorgonio search-and-rescue team for 12 years — died at 2:36 a.m. at Desert Regional Medical Center in Palm Springs, the Riverside County Coroner's Office reported.

He was the second person to die from a head-on collision Saturday on Highway 62 near Yucca Valley.

Calvert was one of five volunteers traveling in a sheriff's Chevy Suburban east on the highway at 5:05 a.m. Saturday when they were struck head-on by Joseph Tidwell, 35, who was driving under the influence of drugs, authorities said.

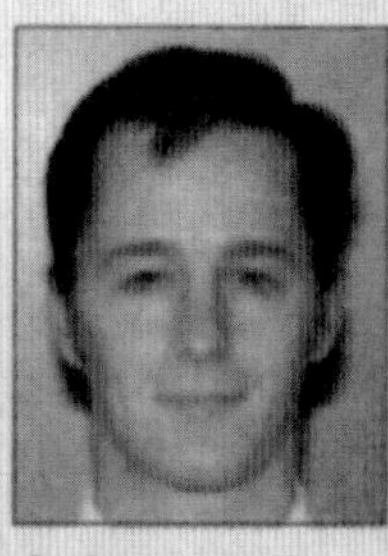

Johnston

The group w
the national par
missing teenage
Diego County.

Search-and-
member Scott J
Redlands died
Team member E
of Yucaipa su

Phil Calvert was riding in the front passenger seat of the search-and-rescue Suburban and took the full force of the head-on collision. Calvert miraculously survived the crash and was taken

worker dies

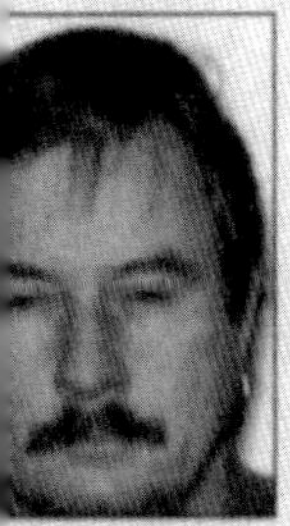

vert

n route to look for a from San

ue team ton, 30, of he scene. 3lanck, 32, ed broken

bones and internal injuries.

His condition was upgraded from critical to good on Tuesday at Arrowhead Regional Medical Center in Colton, said Cindy Beavers of the San Bernardino County Sheriff's Department.

"I was told he was looking forward to having visitors," Beavers said.

The other injured team members were released from hospitals on Sunday, Shane Elliot, 34, of Redlands from Desert Memorial, and David Atchley, 45, from Arrowhead Regional, sheriff's spokesman Chip Patterson said.

Tidwell, a Twentynine Palms resident, is not related to former San Bernardino County Sheriff Floyd Tidwell.

The Twentynine Palms man has a history of DUI-related charges, authorities said.

He was charged Tuesday with two counts of gross vehicular manslaughter and one count of driving under the influence causing injury.

He was arraigned Tuesday in Joshua Tree Superior Court, said acting Supervising Deputy District Attorney Alex Martinez.

See **CRASH**/B6

to Desert Community Hospital where he died of extensive internal injuries two days after the accident. (Article courtesy of *Redlands Daily Facts*.)

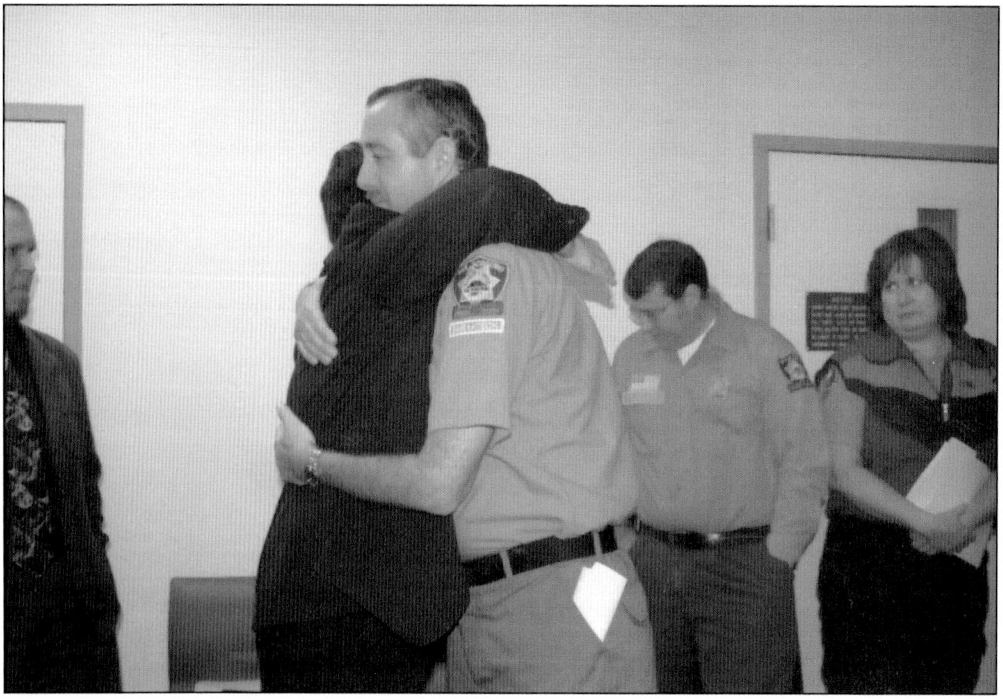

Deputy District Attorney Alex Martinez brought second-degree murder charges against the drunk driver. Many of the team members at the scene of the accident were called to testify. After only 45 minutes of deliberation, the jury found the driver guilty, and he was sentenced to 45 years in prison. Above, Dave Atchley (right), the driver of the team's Suburban, hugs Alex Martinez after the sentencing. Below, Bob Blanck (left) discusses the trial with Martinez. (Photographs courtesy of Jutta Biggerstaff, *Desert Star Newspaper.*)

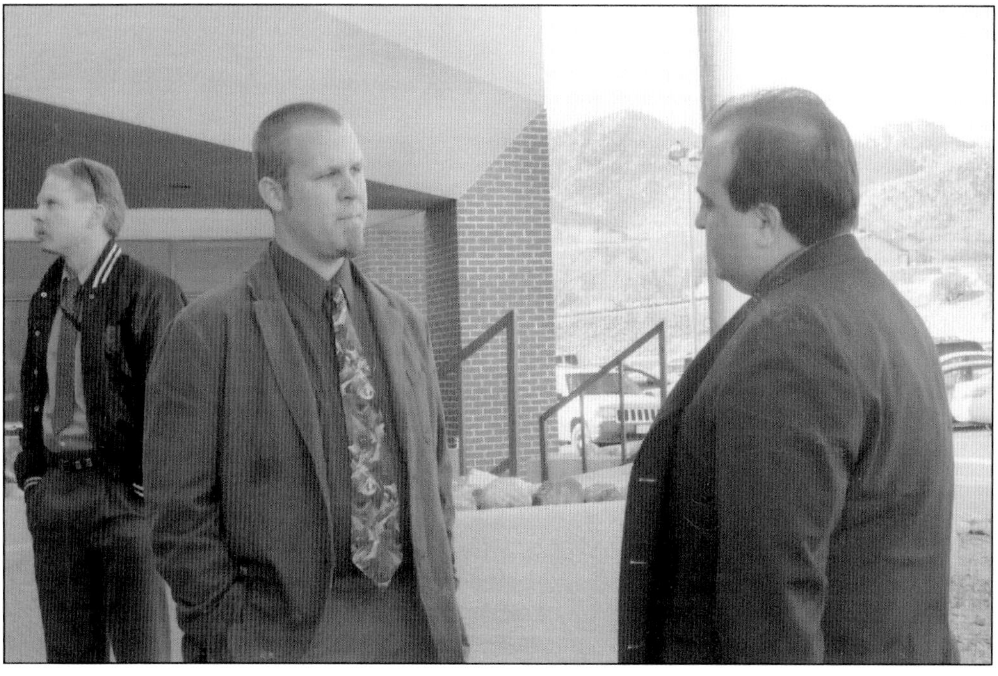

Phil Calvert was like a father to many of the team members. He had been with the team for over 15 years. Calvert was one of the most experienced members of the search-and-rescue group and was an expert in mountaineering, land navigation, ice climbing, rock climbing, and technical rope rescue.

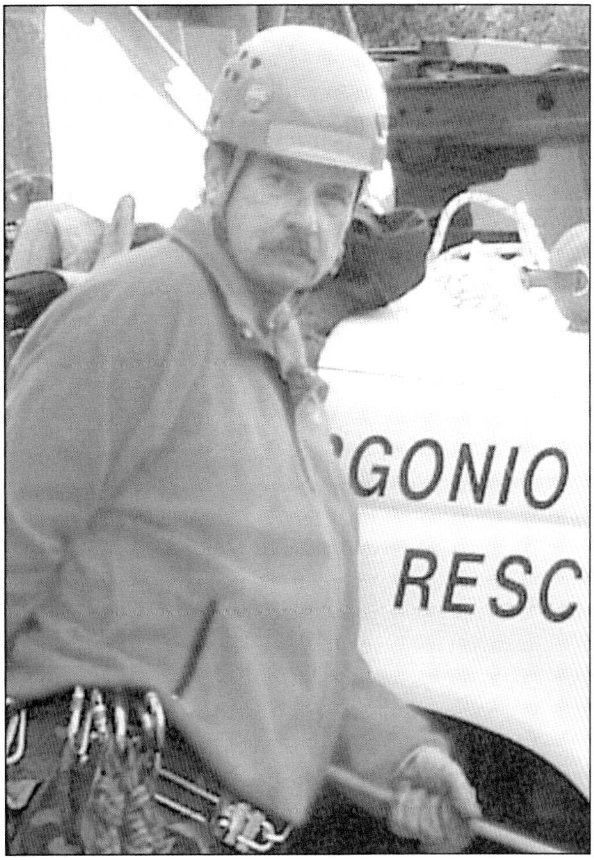

Scott Johnston was a true romantic and had a zeal for life that spilled over to the rest of the team. Although he had only been on the team for six years, he rapidly became one of its most experienced members. He was skilled in mountaineering, snow camping, and technical rope rescue.

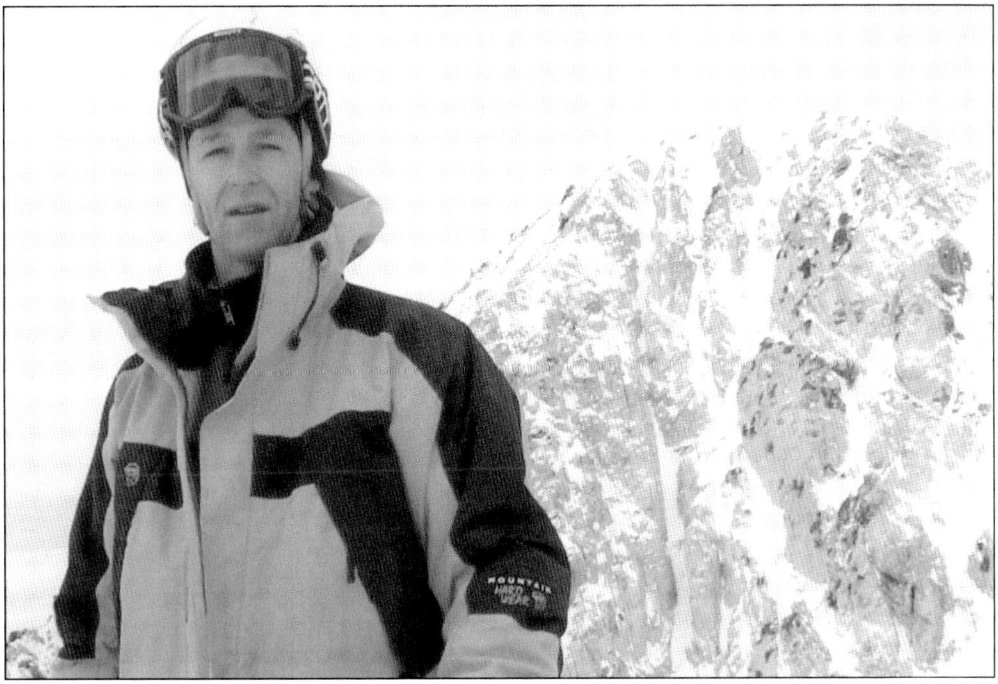

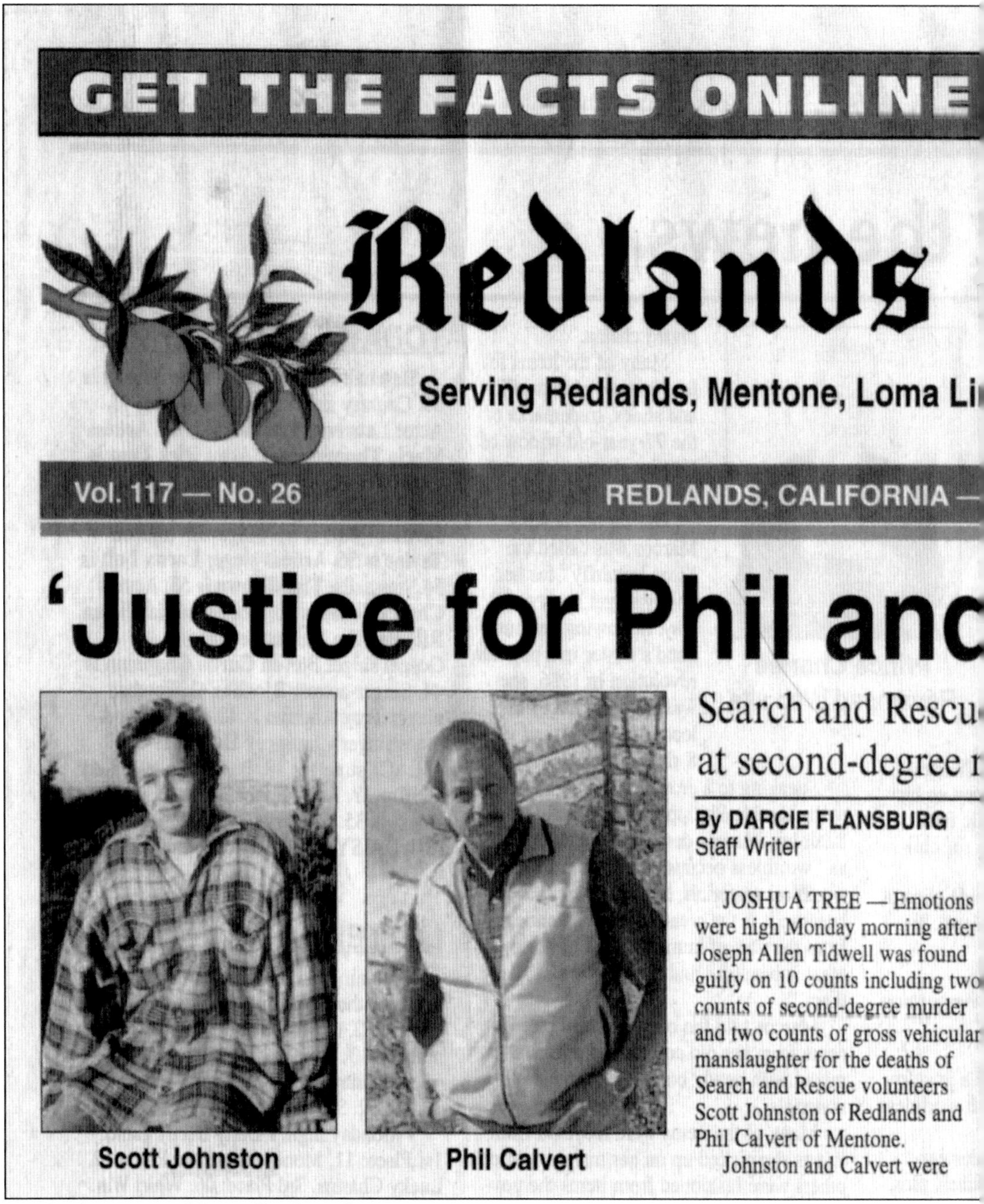
GET THE FACTS ONLINE

Redlands

Serving Redlands, Mentone, Loma Li

Vol. 117 — No. 26 REDLANDS, CALIFORNIA —

'Justice for Phil anc

Search and Rescu

at second-degree r

By DARCIE FLANSBURG
Staff Writer

JOSHUA TREE — Emotions were high Monday morning after Joseph Allen Tidwell was found guilty on 10 counts including two counts of second-degree murder and two counts of gross vehicular manslaughter for the deaths of Search and Rescue volunteers Scott Johnston of Redlands and Phil Calvert of Mentone.

Johnston and Calvert were

On the front page of the *Redlands Daily Facts*, staff writer Darcie Flansberg, who religiously followed the case and the trial and kept Redlands residents up to date on the tragedy the team had faced

www.redlandsdailyfacts.com

Daily Facts

and East Highland. Established 1890

ESDAY, NOVEMBER 21, 2006 25 Cents (Includes CA Sales Tax)

Scott'

lunteers relieved der convictions

iving to Joshua Tree on the orning of July 17, 2004, in earch of a lost hiker when their ar was hit head-on by Tidwell's uck.

Tidwell also was found guilty or possession of heroin, methamhetamines and cocaine, for possession of drug paraphernalia and or possession of syringes without lawful purpose.

"We're just all very relieved," aid Valley of the Falls Search and

See JUSTICE, Page A3

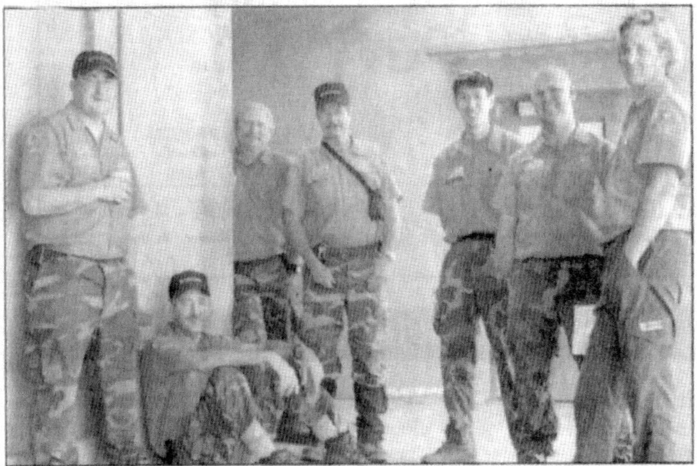

Courtesy photos

Member of the San Gorgonio Search and Rescue team: Shane Eliott (left), Dave Atchley (second to left), Phil Calvert (center), Robert Blanck (second to right) and Scott Johnston (right). Some members of the team are not identified.

over two years earlier, wrote a wonderful article summing up the sentencing of the man who killed Phil Calvert and Scott Johnston. (Article courtesy of *Redlands Daily Facts*.)

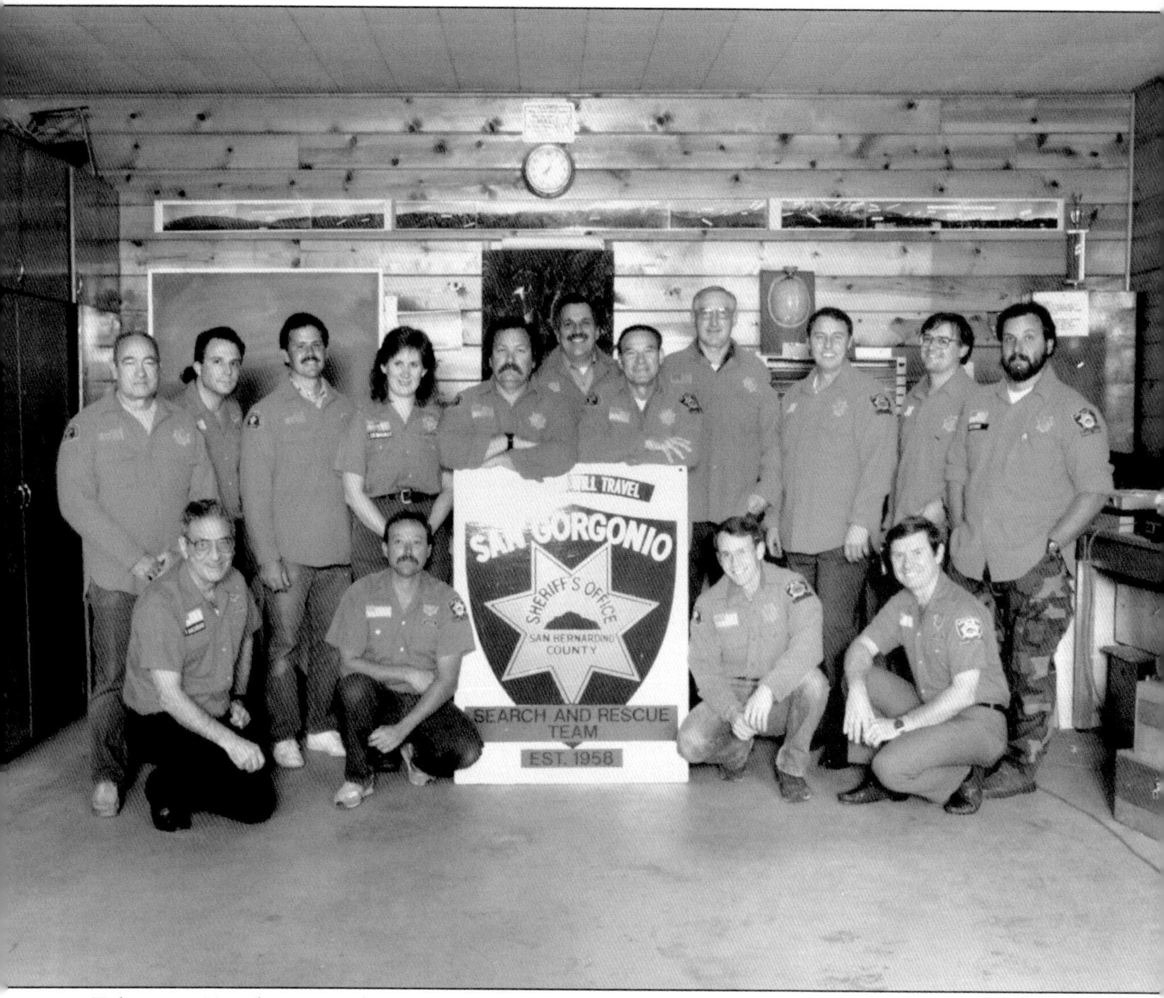

Taken in 1988, this team photograph shows, from left to right, (first row) Dave Farquhar, Richard Bledsoe, Bill Loenhorst, and Jim Lanfried; (second row) Don Walters, Bill Hasper, Frank Ruiz, Ellyn Loenhorst, Skip Brown, Al Stremble, Barry Lee, Packy Wassell, Ken Hurley, Paul Franklin, and Pat Brown.

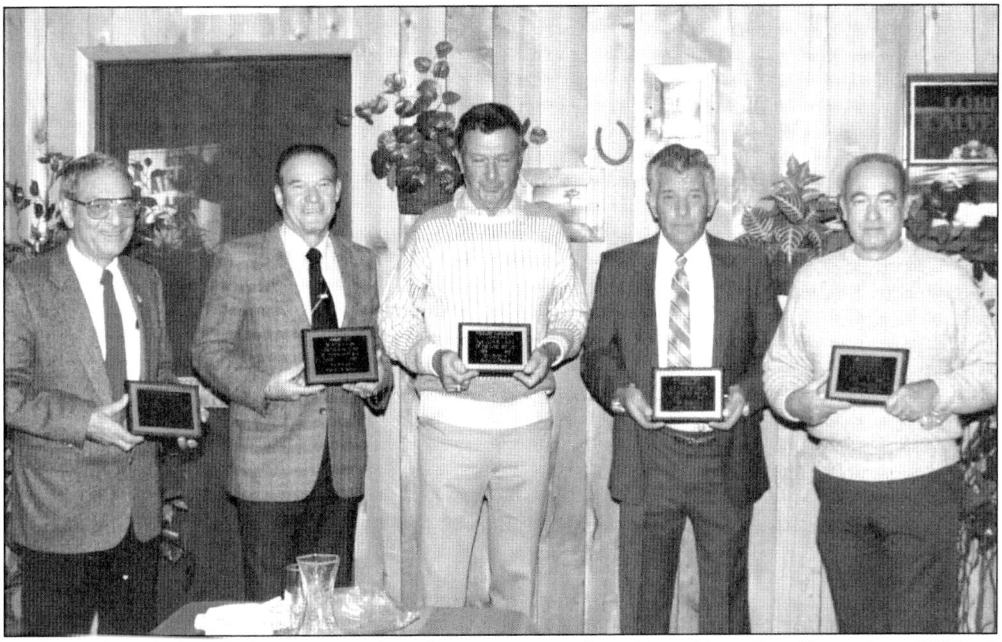

Pictured from left to right in 1998, founding team members Dave Farquhar, Barry Lee, Willard Farquhar, Bill Loenhorst Sr., and Don Walters receive plaques recognizing 40 years of service to San Gorgonio Search and Rescue at the team Christmas party.

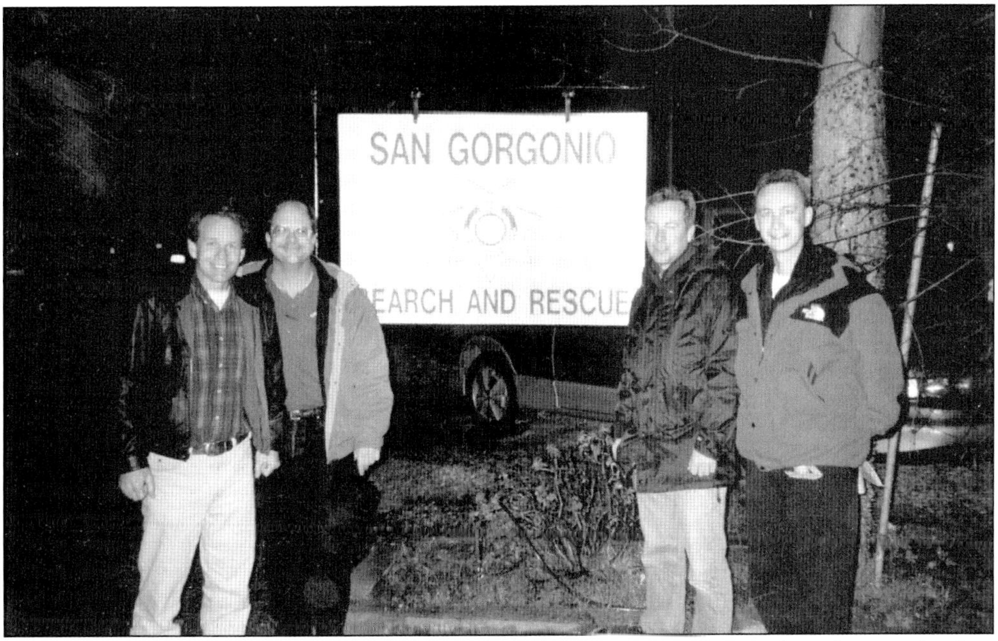

The San Gorgonio Search and Rescue Team is the only one in San Bernardino County that has second-generation members. Following in their fathers' boot tracks are, from left to right, Bill Loenhorst Jr., Bob Lehmann, Clifford Walters, and Rodney Walters. Loenhorst's father served the team as the resident sheriff deputy and coordinator for over 20 years. Lehmann's father served the team for over 20 years, and the Walterses' father serves with the team to this day, with 50 years of service.

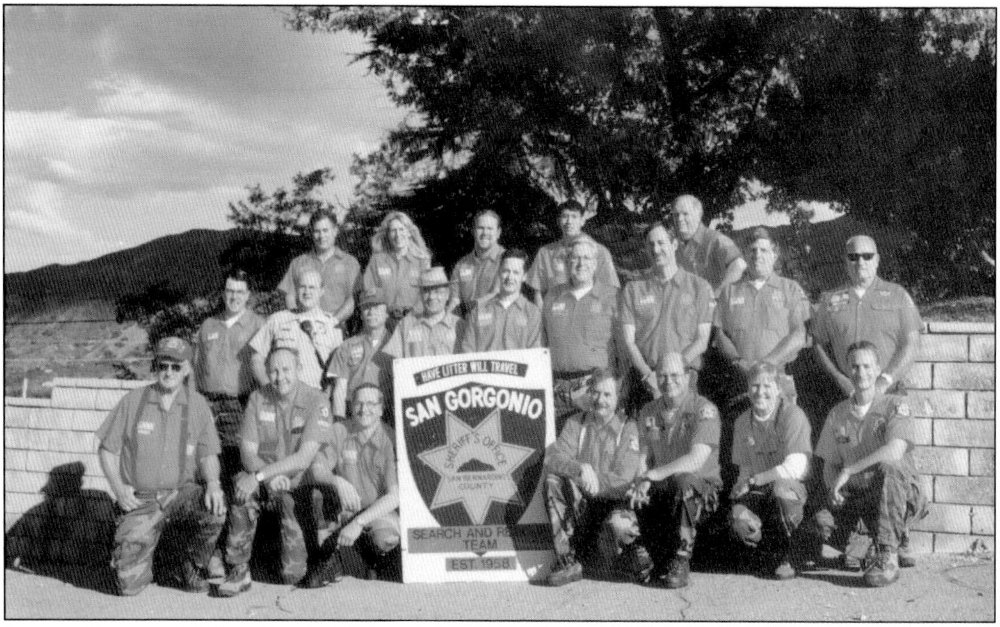

Pictured here in 2000 is the San Gorgonio Search and Rescue Team. From left to right are (first row) Packy Wassell; Paul Butler; Roger Gossett; Phil Calvert; Bob Lehmann; Brenda MacGregor; and Rod Walters; (second row) Deputy Shannon Kovich, team coordinator; Dean Caldwell; Hugh Brown; Tate Stevens; Scott Johnston; Walt Farmer; Dave Atchley; Tom Rutledge; and Don Walters; (third row) Tim Murone; Julie Haggerty; Bob Blanck; Ehren Ngo; and unidentified.

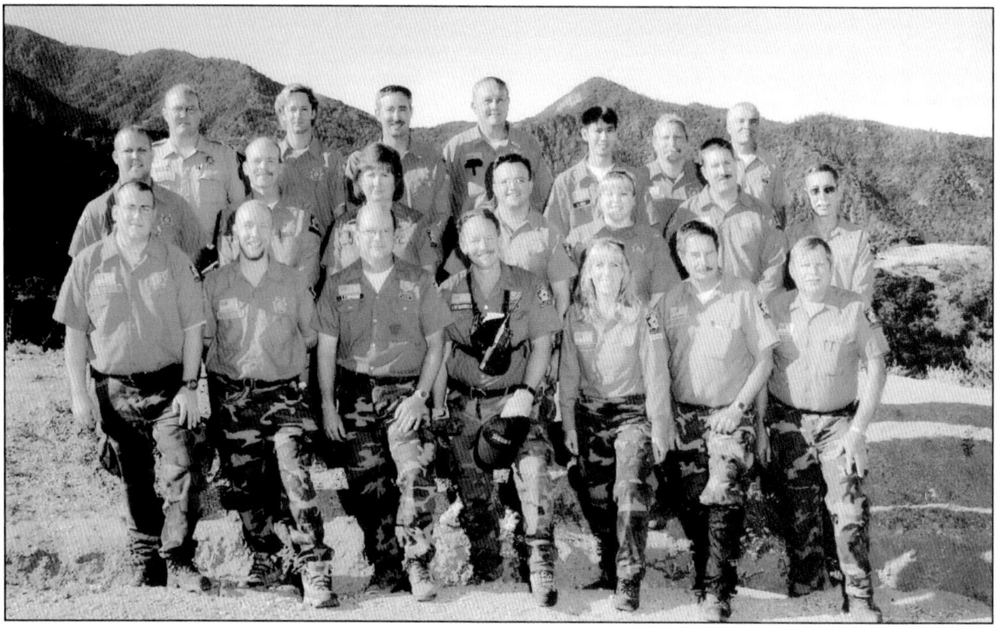

The team poses for this 2004 photograph. From left to right are (first row) Shane Elliot, Bjorn Svenson, Bob Lehmann, Bill Loenhorst, Julie Haggerty, Phil Calvert, and Tom Ziech; (second row) Bob Blanck, Dan Kerr, Ellyn Loenhorst, Lauren Popescue, Jen Coyle, Ned Coyle, and Ron Deitzman; (third row) Deputy Shannon Kovich, Scott Johnston, Dave Atchley, Paul Butler, Ehren Ngo, Kevin Newland, and Hugh Brown.

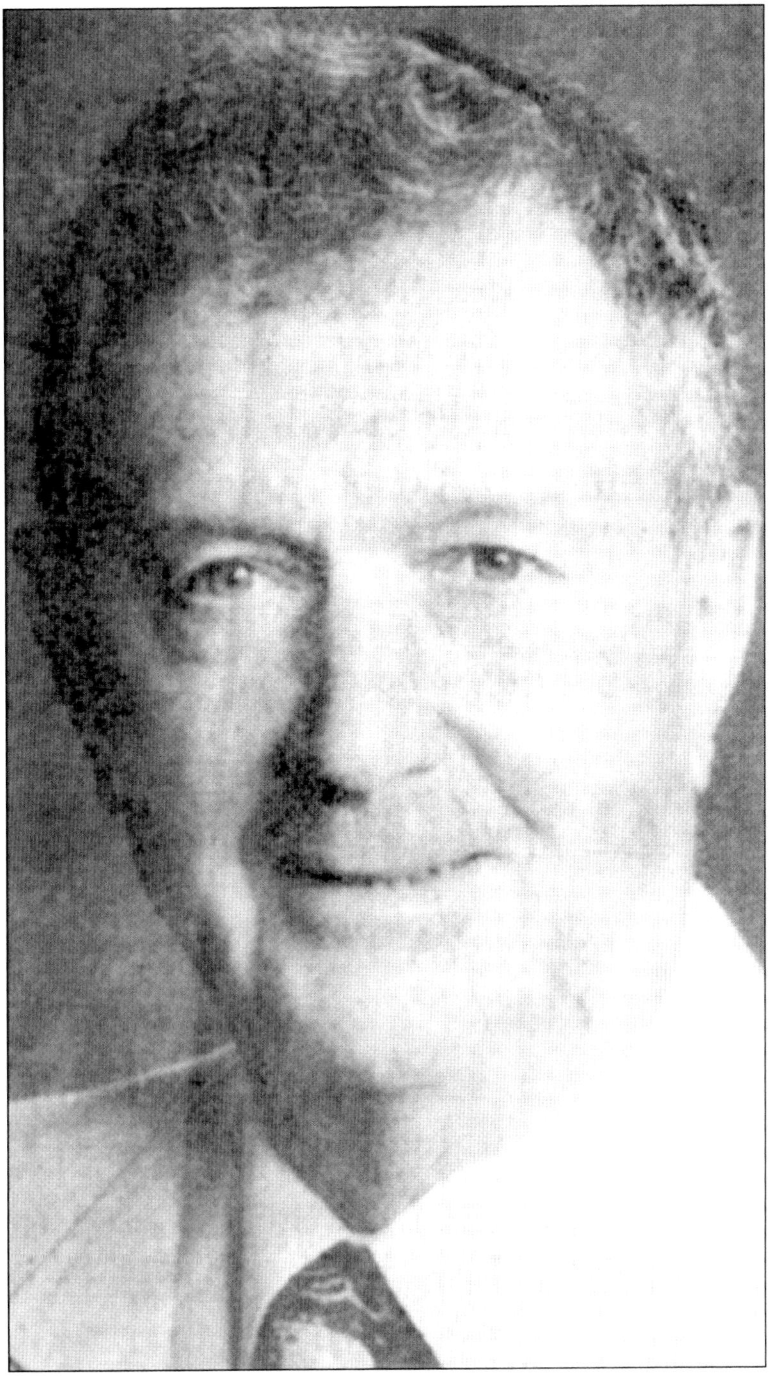

Willard Farquhar, the founder of the San Gorgonio Search and Rescue Team, passed away on April 27, 2002, at 77 years old. Farquhar was with the sheriff's office for 30 years and retired in 1980. He started the team in 1958 to search for the lost and rescue the injured in the San Gorgonio Wilderness. Willard truly loved the team. From this humble beginning, San Bernardino County now has one of the largest and most sophisticated search-and-rescue team systems of any law enforcement agency in the United States.